# Further Practical Microelectronics

**Units in this series**

| | |
|---|---|
| Microelectronic Systems | Level I |
| Microelectronic Systems | Level II |
| Microelectronic Systems | Level III |
| Microprocessor-based Systems | Level IV |
| Microprocessor-based Systems | Level V |
| Microprocessor Appreciation | Level III |
| Microprocessor Principles | Level IV |

**Practical assignments in this series**

An Introduction to Practical Microelectronics
Further Practical Microelectronics

**Study guides in this series**

TEC Study Guide for Microelectronic Systems Level I
TEC Study Guide for Microelectronic Systems Level II
TEC Study Guide for Microelectronic Systems Level III
TEC Study Guide for Microprocessor-based Systems Level IV
TEC Study Guide for Microprocessor-based Systems Level V
TEC Study Guide for Microprocessor Appreciation Level III
TEC Study Guide for Microprocessor Principles Level IV

# Further Practical Microelectronics

## Peter Simmons
*Brighton Technical College*

TECHNICIAN EDUCATION COUNCIL
in association with
HUTCHINSON
*London Melbourne Sydney Auckland Johannesburg*

Hutchinson & Co. (Publishers) Ltd
An imprint of the Hutchinson Publishing Group
17–21 Conway Street, London W1P 6JD

Hutchinson Publishing Group (Australia) Pty Ltd
PO Box 496, 16–22 Church Street,
Hawthorne, Melbourne, Victoria 3122

Hutchinson Group (NZ) Ltd
32–34 View Road, PO Box 40–086, Glenfield, Auckland 10

Hutchinson Group (SA) (Pty) Ltd
PO Box 337, Bergvlei 2012, South Africa

First published 1984

© Technician Education Council 1984

Set in Times by Grainger Photosetting, Southend-on-Sea, Essex, England

Printed and bound in Great Britain by
Anchor Brendon Ltd,
Tiptree, Essex

**British Library Cataloguing in Publication Data**
Technician Education Council
  Further practical microelectronics.
  1. Microelectronics
  I. Title
  621.381'71    TK7874

ISBN 0 09 151671 4

# Contents

# Preface

This book is one of a series on microelectronics/microprocessors published by Hutchinson on behalf of the Technician Education Council. The books in the series are designed for use with units associated with Technician Education Council programmes.

In June 1978 the United Kingdom Prime Minister expressed anxiety about the effect to be expected from the introduction of microprocessors on the pattern of employment in specific industries. From this stemmed an initiative through the Department of Industry and the National Enterprise Board to encourage the use and development of microprocessor technology.

An important aspect of such a development programme was seen as being the education and training of personnel for both the research, development and manufacture of microelectronics material and equipment, and the application of these in other industries. In 1979 a project was established by the Technician Education Council for the development of technician education programme units (a unit is a specification of the objectives to be attained by a student) and associated learning packages, this project being funded by the Department of Industry and managed on their behalf by the National Computing Centre Ltd.

TEC established a committee involving industry, both as producers and users of microelectronics, and educationists. In addition widespread consultations took place. Programme units were developed for technicians and technician engineers concerned with the design, manufacture and servicing aspects incorporating microelectronic devices. Five units were produced:

| | |
|---|---|
| Microelectronic Systems | Level I |
| Microelectronic Systems | Level II |
| Microelectronic Systems | Level III |
| Microprocessor-based Systems | Level IV |
| Microprocessor-based Systems | Level V |

Units were also produced for those technicians who required a general understanding of the range of applications of microelectronic devices and their potential:

| | |
|---|---|
| Microprocessor Appreciation | Level III |
| Microprocessor Principles | Level IV |

This phase was then followed by the development of the learning packages, involving three writing teams, the key people in these teams being:

Microelectronic Systems I, II, III – P. Cooke
Microprocessor-based Systems IV – A. Potton
Microprocessor-based Systems V – M. J. Morse
Microprocessor Appreciation III – G. Martin
Microprocessor Principles IV      – G. Martin

The project director during the unit specification stage was N. Bonnett, assisted by R. Bertie. Mr Bonnett continued as consultant during the writing stage. The project manager was W. Bolton, assisted by K. Snape.

# Acknowledgements

The author and publisher are grateful to the following for their permission to reproduce illustrations, and for help in providing material:

Figure 1.5: photograph and technical details supplied by Tektronix Inc; Figure 3.5: Pro-log Corporation of Monterey, California, material supplied by Technitron Inc (UK branch) as UK agents for Pro-log; Figures 3.2, 4.33, 6.3, 6.4, 6.5: R. S. Components Ltd; Figure 4.1: The Open University; Figures 4.15, 4.16: Hewlett Packard Ltd; Figure 5.18: Motorola Ltd; Data sheets in Appendix 2: Intel Corporation, Ferranti Electronics Ltd, R. S. Components Ltd.

Every effort has been made to reach copyright holders, but the publisher would be grateful to hear from any source whose copyright they may unwittingly have infringed.

# Introduction

This book has been written as an extension of *Introduction to Practical Microelectronics*. It contains practical exercises relating to the TEC units *Microprocessor-based Systems Level IV* and *Microprocessor-based Systems Level V*.

Because it is written as part of a series it is not completely self-contained; it is assumed that readers have some familiarity with the material developed in earlier volumes and will have access to the manual written for the particular development system with which they are working.

*Peter Simmons*

# Chapter 1 Introduction to the use of development systems

*Objectives of this chapter*    *When you have completed this chapter you should be able to:*

1   *Understand the uses of development systems.*
2   *State that the development system is used to assist:*
    *(a)*   *Development of program flow chart(s)*
    *(b)*   *Writing of program*
    *(c)*   *Testing of program*
    *(d)*   *Modification, where necessary*
    *(e)*   *Simulation/emulation of system*
    *(f)*   *Further modification, where necessary.*
3   *State that typical facilities within a development system can be:*
    *(a)*   *Operating program consisting of:*
       *(i)*     *Operating system and monitor*
       *(ii)*    *Editor*
       *(iii)*   *Assembler*
       *(iv)*   *Debugger*
       *(v)*    *PROM programmer*
       *(vi)*   *Loader*
       *(vii)*   *Linker*
       *(viii)*   *Locater*
       *(ix)*   *Software trace*
    *(b)*   *Memory for:*
       *(i)*     *Applications program development*
       *(ii)*    *Operating system*
       *(iii)*   *Non-volatile bulk storage*
    *(c)*   *Keyboard and VDU/printer*
    *(d)*   *PROM programmer*
    *(e)*   *I/O facilities*
    *(f)*   *In-circuit emulator*
4   *Use a development system to enter, edit, assemble and debug programs.*
5   *Test software with appropriate hardware peripherals.*
6   *Given a flow chart, instruction set and peripheral hardware, use a system to demonstrate that a program will perform a specified task.*

## 1.1 Introduction

The development of a microprocessor-based logic control system follows a series of defined steps. These are summarized in the flow

Figure 1.1 *The stages in the development of a microprocessor-based system*

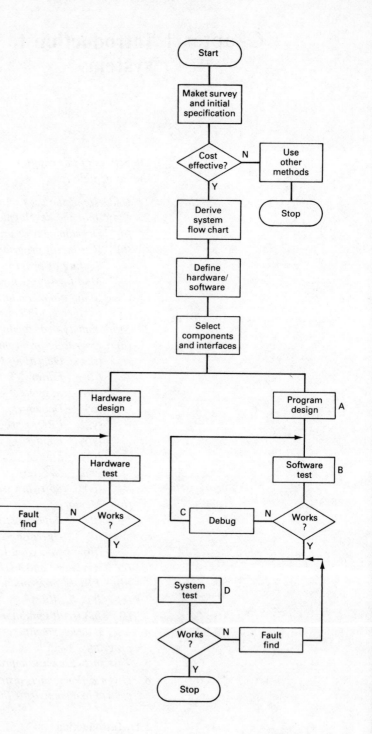

chart of Figure 1.1. A development system is designed to speed the operations in the processes labelled A, B and C in the figure.

A development system is a design tool used to develop and test the software which will be incorporated into a microprocessor-based product. Some advanced systems allow the testing of complete hardware units (process D in Figure 1.1) using the techniques of in-circuit emulation (ICE) and real-time tracing of the program.

As has been explained in *An Introduction to Practical Microelectronics*, and *Microelectronics Systems Level I-III*, both data and instructions need to be presented as binary numbers before the processor can 'understand' them. Development systems provide mechanisms for generating these binary programs, amending them, tracing errors in them, etc; the more sophisticated systems offer a greater range of facilities and greater versatility than the smaller (cheaper) systems. Larger systems would also normally offer facilities whereby a program once developed, could be transferred to a PROM (programmable read-only memory) or EPROM (erasable PROM) device which could then be removed and incorporated in a stand-alone microprocessor system designed to perform a specified function.

The main function of a development system is thus to simplify the task of the programmer or development engineer.

### 1.2 Essential features of a development system

Whilst it is possible to debate at length the minimum range of facilities a development system should have, there will always be two elements:

1 An editor       2 An assembler

The task of the editor is to facilitate the entry of data and instructions to the system. It will provide a range of screen prompts, indicating for example when one task is completed and the next may be started, or telling the operator to wait until a block of data or a program has been loaded from tape or disc. It will also control the flow of information from the keyboard into memory. Most editors will provide mechanisms for renumbering lines, deleting lines, changing lines or parts of lines, and inserting new lines in a program.

The assembler is itself a program and it has the task of translating the program statements written by the operator into machine code (binary words) ready for execution by the processor. The assembler is, in comparison with high-level programming languages like BASIC or FORTRAN restricted; for example, there must be a one-to-one correspondence between a command input by an operator and a machine code instruction. This is in contrast to one instruction in a high-level language which may require many machine code instructions for its execution. The operator must present these instructions in a specified way, usually using mnemonics (often three-letter abbreviations) for instructions, and specifying addresses and variables in hexadecimal notation.

Any small computer system operating with a TV as a display unit could be programmed to offer the two facilities of editor and assembler, with the assembled machine (or object) code program being written as an array to a specified memory address which could then be run as a machine code program. The facilities that can be provided in such a simple system are restricted only by the ingenuity of the programmer.

Systems which support an editor invariably require information to be presented in a specified format, with the different parts of an instruction separated by a particular character such as a full stop or colon. The assembler may require programs to be presented as three-letter mnemonics, or possibly as a hexadecimal code representing each instruction, and will convert these to machine code. More sophisticated systems allow development work for a number of different microprocessors (typically 8080, Z80, M6800), accepting the mnemonics appropriate to that which is selected and generating the correct object code. Simpler systems are specific to a single microprocessor.

## 1.3   A simple development system – HEKTOR

HEKTOR is the name used for a simple development system developed primarily for instructional purposes. It has been described in detail in *An Introduction to Practical Microelectronics*.

For the HEKTOR each line of source program is divided into four fields.

LABEL: OPERATION, OPERAND; COMMENT

Note that the LABEL field ends with a colon and the COMMENT field is preceded by a semicolon.

For example, consider the following source program:

```
START: LXI SP, 3FFFH; SET UP USER STACK
       LXI H,   3800H; DATA ADDRESS

TOTAL: MOV A, M     ; 1ST NUMBER
       ORA A        ; CLEAR CARRY
       INX H        ; MOVE POINTER
       ADC M        ; ADD
       RAR          ; DIVIDE BY 2
       INX H        ; MOVE POINTER
       MOV M, A     ; SAVE ANSWER
STOP:  JMP 00 57H   ; RETURN TO MONITOR
```

Before this source program can be loaded into the computer it is necessary to specify that the editor is to be used.

On the HEKTOR this is done by typing E followed by a space and the return button. Although this 'calls' the editor program, and this is confirmed by the screen prompt *EDITOR*, it is still necessary to specify what is required of the editor; in other words, it needs to be told that program lines are to be entered. On the HEKTOR this is done by typing I (for Insert new line) followed by a space and return. The computer responds by prompting with 0001, the number of the first program line, and the program can now be typed in, line by line.

The HEKTOR editor has a number of useful features.

1   Whilst entering a line, a cursor key ← can be used to step backwards, deleting characters as it goes. You can then retype from that point. This is of obvious value in correcting errors.
2   Any program line can be inspected by typing P (Print) followed by the line number of the line you want displayed. Any set of lines can be inspected by typing, P*x,y* where, if *x* and *y* are valid line numbers, lines *x* to *y* inclusive will be displayed.
3   In a complete program listing, any character can be altered by positioning the cursor (an underline) at the character position just after the error. Using the backspace arrow followed by the required character clears the error.
4   New lines can be inserted. The following is an example of this procedure.

Figure 1.2   *A source program*

```
0001 START: LXI SP,3FFH ;SET UP USER STACK
0002  LXI H, 3800H ;DATA ADRESS
0003 TOTAL: MOV A, M ;1ST NUMBER
0004  ORA A ;CLEAR CARRY
0005  INX H ;MOVE POINTER
0006  ADC M ;ADD
0007 RAR ;DIVIDE BY TWO
0008  INX H ;MOVE POINTER
0009  MOV M,A ;SAVE ANSWER
0010 STOP: JMP 0057H ;RETURN TO MONITOR
```

Figure 1.2 shows a source listing. The source text must include certain instructions to the assembler. In particular, the assembler must be informed where the source program finishes, otherwise it may well try to assemble the garbage in RAM. This and similar instructions are known as *Assembler directives*. Two such directives are ORG and END. The source listing in Figure 1.2 is missing these directives.

ORG informs the assembler of the memory address of the start of the object code which it will generate. In the case of this program it is to be stored in memory starting at address 3100H. So, for this program, the statement:

ORG 3100H

must be included at the start.

Typing I, space, 1 followed by RETURN on the HEKTOR allows a new line to be inserted above the line which has as its line number 1. All subsequent lines will be automatically renumbered as can be seen by 'printing' the new complete program which now contains 11 lines. This can be done to enter the ORG directive.

END must always be included as the last statement in the program. It informs the assembler that any further text in the workspace memory should be ignored.

Typing I, space, 12 followed by RETURN would permit a new line 12 to be inserted. This will be needed to enter the END directive.

The final listing is shown in Figure 1.3.

Figure 1.3  *A source program with assemble directives*

```
0001  ORG 3100H
0002 START: LXI SP,3FFFH ;SET UP STACK
0003  LXI H, 3800H ;DATA ADDRESS
0004 TOTAL: MOV A,M ;FIRST NUMBER
0005  ORA A ;CLEAR CARRY
0006  INX H  ;POINT TO NEXT NUMBER
0007  ADC M ;ADD NUMBER
0008  RAR ;DIVIDE BY 2
0009  INX H ;MOVE POINTER
0010  MOV M,A ;SAVE ANSWER
0011 STOP: JMP 0057H ;RETURN TO MONITOR
0012 END
```

5   The HEKTOR editor also enables the user to save and verify the programs on to tape and also to load them back into the computer.

The assembler program offered by the HEKTOR computer has four main features:

1   It translates source programs into object programs.
2   It assigns actual memory addresses to the symbolic addresses used by the programmer.
3   It assigns addresses for the program.
4   It will check for syntax errors and inform the programmer if any are found.

*It will not detect logic errors* in the program – this is a job for the programmer.

The assembler produces, in addition to the object code, a listing file

Figure 1.4　*An assembler listing of*
*the program of Figure 1.3*

```
                    PASS 1

                    PASS 2

                    PASS 3

3100        0001            ORG  3100H
3100 31FF3F  0002 START:    LXI  SP,3FFFH  ;SET UP STACK
3103 210038  0003           LXI  H,3800H   ;DATA ADDRESS
3106 7E      0004 TOTAL:    MOV  A,M       ;FIRST NUMBER
3107 B7      0005           ORA  A         ;CLEAR CARRY
3108 23      0006           INX  H         ;POINT TO NEXT NUMBER ???
3109 8E      0007           ADC  M         ;ADD NUMBER
310A 1F      0008           RAR            ;DIVIDE BY 2
310B 23      0009           INX  H         ;MOVE POINTER
310C 77      0010           MOV  M,A       ;SAVE ANSWER
310D C35700  0011 STOP:     JMP  0057H     ;RETURN TO MONITOR
3110         0012           END
```

which joins the source program and object code in a special format which makes the program more readable. An example of a listing program is shown in Figure 1.4.

## 1.4　A more sophisticated development system

A number of more sophisticated development systems are currently available. The exact features offered by each vary from one to another but we will explore the facilities typically available by considering in more detail the Tektronix 8550 microprocessor lab. (see Figure 1.5). It has all the facilities of HEKTOR, and in addition has a wider range of hardware facilities and software 'utilities'.

Figure 1.5　*The Tektronix 8550, a*
*typical modern microcomputer*
*development system supporting a wide*
*range of eight and sixteen bit*
*microprocessors*

It is a disc-based system; that is, programs are stored on 8 inch floppy discs instead of cassette tapes and each program is provided with a *file name* to aid identification and retrieval.

A more powerful editor and assembler are provided with a debug and trace facility. The 8550 also provides for EPROM programming and in-circuit emulation.

In this development system the assembler can be changed to suit different processors, i.e. 8085, 6800, Z80, etc, whereas HEKTOR could only assemble 8085 programs. Each of these facilities is now briefly examined.

### 8550 microprocessor lab. hardware components

The 8550 microprocessor lab. internal architecture centres around a system microprocessor that uses other microprocessors to perform different software and hardware support functions. The system contains 16 Kbytes of system random-access memory (RAM) and up to 64 Kbytes of RAM program memory (depending on the options

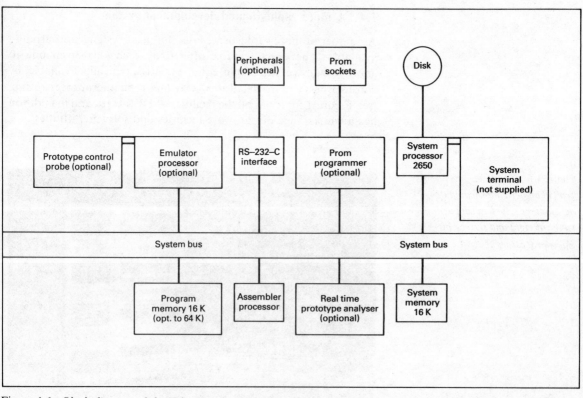

Figure 1.6   *Block diagram of the Tektronix development system*

selected). The system also supports two flexible disc drives with approximately 315 Kbytes on each disc.

An 8550 microprocessor lab. system block diagram is shown in Figure 1.6. The system contains three microprocessors – the system processor, the assembler processor and the emulator processor. Each microprocessor resides on a separate plug-in circuit card in the system mainframe. These cards are connected to each other through a common system bus. Also residing in the mainframe is the optional PROM programmer, the RS–232–C interface with three I/O ports, the 16 Kbyte system memory and the standard 16 Kbyte program memory (expandable to 64K).

The flexible disc unit is housed in a separate chassis and communicates with the other system components through the system processor. Other optional system peripherals such as the CT8100 CRT terminal and the LP8200 line printer communicate with the system through the RS–232–C interface.

The following is a brief description of each component in the system.

*System processor*
The system processor performs the following supervisory functions:

1  *System input/output*  Directs all I/O activity for the system peripherals, such as the flexible disc, the console and the line printer.
2  *File management*  Organizes, stores and retrieves user programs and system programs from the disc drives.
3  *Text editing*  Executes the text editor program and maintains text files on the flexible disc unit.
4  *Debugging*  Executes the debug program and controls the emulator processor through separate debug hardware.
5  *System utilities*  Performs all system utility functions such as processing the messages between system peripheral devices.
6  *PROM programming*  Monitors and controls all PROM activity.

*Assembler processor*
The assembler processor runs the Tektronix assembler program when the TEKDOS ASM command is executed. All assembler I/O activity to and from the flexible disc unit is handled by the system processor.

*Emulator processor*
The emulator processor, a system option, runs and debugs user programs written for a particular microprocessor. A separate processor is available for each microprocessor you wish to emulate.

The emulator processor serves two purposes. First, the emulator processor runs the user program whilst the system debugger program is active. This detects program run-time errors and program logic errors. Second, with the addition of an optional prototype control probe, the emulator processor takes the place of the actual microprocessor in the prototype under development. The user program can then drive and test the prototype hardware while under the supervision of the debug system.

### System memory

The system memory is a 16 Kbyte dynamic RAM located on a separate module within the main chassis. The system memory is accessed only by the system processor and is used to store TEKDOS operating system programs while they are executing. The system memory also provides buffer space for all I/O activities.

### Program memory

The standard 16 Kbyte program memory is located on a separate module within the main chassis. Additional 16 Kbyte memory modules can be added to increase the total capacity to 64 Kbytes. The primary purpose of program memory is to store a user program while the program is being executed by the emulator processor. The system processor also uses program memory as a text buffer during text editing sessions.

### Prototype control probe

The optional prototype control probe consists of cables, interface circuits and a 40-pin connector. The connector plugs into the empty microprocessor socket on the prototype circuit board. The prototype control probe allows the emulator processor and program memory to take the place of the actual microprocessor and its associated memory in the prototype. Thus the user program can be run, tested and debugged in the prototype while under the supervision of the debug system.

The following three emulator operational modes are available with the prototype control probe plugged into the prototype:

1   *System mode (mode 0)*   The emulator processor runs the program residing in program memory.
2   *Partial emulation mode (mode 1)*   The emulator processor runs the program residing in program memory and prototype memory. All I/O signals and data are supplied by the external prototype hardware.
3   *Full emulation mode (mode 2)*   The emulator processor runs the program resident in the external prototype memory. All I/O signals and data are also supplied by the prototype hardware.

### Real-time prototype analyser

The optional real-time prototype analyser enables you to

dynamically monitor the prototype address bus, data bus and up to eight other locations of your choice on the prototype circuit board.

The analyser's main function is to locate critical timing problems and hardware/software sequence problems in the prototype during the last stages of system integration and debugging. The analyser monitors prototype activity while the prototype is running at full speed. The test results are printed on either the system console or the optional line printer.

*PROM programmer*
The PROM programmer option allows user programs to be transferred from program memory into PROM chips. These PROM chips are then plugged into the prototype memory sockets and provide permanent program instructions for the prototype micro-processor. Not only can user programs be transferred from program memory into PROMs, but the reverse action can also take place – the contents of PROMs can be read into program memory. In addition, the user program residing in a PROM can be compared with the user program residing in program memory. The differences are displayed on the system console. This comparison technique is used to verify the contents of a PROM.

*RS–232–C interface*
The RS–232–C interface board provides three I/O ports for connecting optional peripheral devices to the system. Any device that conforms to the EIA standard RS–232–C can be connected to the interface board. Typically, devices such as the LP8200 line printer are connected to the interface. A larger host computer can also be transferred from the host and down-loaded into program memory for execution.

*Flexible disc unit*
A flexible disc unit is the on-line mass storage device for the 8550 microprocessor lab. system. The flexible disc unit consists of two separate disc drive assemblies, a microprocessor controller, a power supply and a cabinet. The flexible disc unit communicates directly with the system processor module through an interconnecting cable. Another flexible disc unit can be connected into the system to provide a four-disc drive option.

Before programs or data can be stored on a floppy disc it is necessary to 'write enable' it as shown in Figure 1.7. In most systems one disc stores all the operating programs such as the editor, assembler and debugger etc. and is known as the *system disc*. The other disc is used to store 'user programs' and data. It is generally called the *user disc*. System discs are kept 'write protected' by removing the label so that expensive system software is not erased or corrupted accidentally.

Figure 1.7   *'Write enabling' an 8 inch floppy disc*

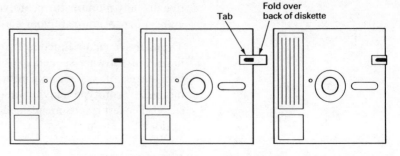

Figure 1.7 shows the 'write enabling' of an 8 inch floppy disc.

*System Terminal*
The 8550 microprocessor lab. system terminal serves as the main communication channel between the system and the operator. (The system terminal is also referred to as the system console in these notes.)

Any terminal-like device can be used as the system terminal if the device has a keyboard, a display and an RS–232–C communications port. The terminal cable is connected directly to the system processor board in the mainframe.

**Operating system**

The operation of the microprocessor lab. is under the control of a set of programs known as the disc-operating system or DOS. In this system it is known as TEKDOS and provides the system user with the utility programs for microprocessor development. Some of these are now examined.

*Text editor*
The text editor is invoked by the TEKDOS EDIT command and performs powerful text editing functions. The text editor is used to:

1   Enter new user programs into memory, then store the programs on disc.
2   Correct user programs for errors detected during assembly.

The text editor can also be used to store and update the support documentation for the prototype under development. Complete text editor instructions are fully described later.

*Tektronix assembler*
After a source program has been entered and stored on a flexible disc unit by the text editor, the user program must be translated into machine-executable object code. This function is performed by the

Tektronix assembler. The assembler then stores the assembled object code on disc in another file.

The assembler is loaded from disc into program memory and runs on the assembler processor. The assembler uses free space in program memory for I/O buffers and symbol tables. Versions of the Tektronix assembler exist for each microprocessor supported by the 8550 microprocessor lab. A separate disc is used for each version.

*Linker*
The linker software is considered a submodule of the assembler software and is provided with each system disc. The linker is used to join several smaller user program modules into one large program. This feature allows several software engineers to work on program segments independently and then join the segments into a larger workable program.

*Emulator*
The emulator software allows user programs to be loaded into the optional emulator processor for operating, testing and debugging.

**Debug System**

Since the assembler software only detects syntax errors in the user program, a number of program logic errors usually remain undetected until the user program is executed. The debug system monitors user program execution on the emulator processor and the prototype microprocessor. The debug software allows you to examine, trace and modify the program.

The program can then be tested under the control of the DEBUG program. Figure 1.8 shows a typical trace produced from DEBUG. Notice that this is a complete trace table, not the output produced by HEKTOR in its single-step mode

So far the operation of the system has been similar to HEKTOR except for the added speed of saving and retrieving programs and greater flexibility of the system commands. There are, however, a number of other facilities which recommend the more advanced system. These are as follows:

*Library programs*
Owing to the expanded storage facilities of the floppy disc, it is possible to set up a library of standard and/or often-used routines which can be added to an object code program by a linker program. This can simplify source programs by making them shorter and hence easier to follow. It is important, though, that the preamble informs the reader what these library programs do.

Figure 1.8   *A trace output from the Tektronix debugger*

```
OC    INST    MNEM OPER     SP   RF RA RB RC RD RE RH RL   IX    IY
2000  3E0F    LD   A,0F     FFFC 11 0F 06 32 FF FF 01 36  0000  0000
2002  D382    OUT  (82),A   FFFC 11 0F 06 32 FF FF 01 36  0000  0000
2004  0E32    LD   C,32     FFFC 11 0F 06 32 FF FF 01 36  0000  0000
2006  0608    LD   B,08     FFFC 11 0F 08 32 FF FF 01 36  0000  0000
2008  212620  LD   HL,2026  FFFC 11 0F 08 32 FF FF 20 26  0000  0000
200B  7E      LD   A,(HL)   FFFC 11 03 08 32 FF FF 20 26  0000  0000
200C  D380    OUT  (80),A   FFFC 11 03 08 32 FF FF 20 26  0000  0000
200E  CD1A20  CALL 201A     FFFA 11 03 08 32 FF FF 20 26  0000  0000
201A  E5      PUSH HL       FFF8 11 03 08 32 FF FF 20 26  0000  0000
201B  11FFFF  LD   DE,FFFF  FFF8 11 03 08 32 FF FF 20 26  0000  0000
201E  210003  LD   HL,0300  FFF8 11 03 08 32 FF FF 03 00  0000  0000
2021  19      ADD  HL,DE    FFF8 11 03 08 32 FF FF 02 FF  0000  0000
2022  38FD    JR   C,FF     FFF8 11 03 08 32 FF FF 02 FF  0000  0000
2021  19      ADD  HL,DE    FFF8 11 03 08 32 FF FF 02 FE  0000  0000
2022  38FD    JR   C,FF     FFF8 11 03 08 32 FF FF 02 FE  0000  0000
2021  19      ADD  HL,DE    FFF8 11 03 08 32 FF FF 02 FD  0000  0000
2022  38FD    JR   C,FF     FFF8 11 03 08 32 FF FF 02 FD  0000  0000
2021  19      ADD  HL,DE    FFF8 11 03 08 32 FF FF 02 FC  0000  0000
2022  38FD    JR   C,FF     FFF8 11 03 08 32 FF FF 02 FC  0000  0000
2021  19      ADD  HL,DE    FFF8 11 03 08 32 FF FF 02 FB  0000  0000
2022  38FD    JR   C,FF     FFF8 11 03 08 32 FF FF 02 FB  0000  0000
2021  19      ADD  HL,DE    FFF8 11 03 08 32 FF FF 02 FA  0000  0000

OC    INST    MNEM OPER     SP   RF RA RB RC RD RE RH RL   IX    IY
2022  38FD    JR   C,FF     FFF8 11 03 08 32 FF FF 02 FA  0000  0000
2021  19      ADD  HL,DE    FFF8 11 03 08 32 FF FF 02 F9  0000  0000
2022  38FD    JR   C,FF     FFF8 11 03 08 32 FF FF 02 F9  0000  0000
2021  19      ADD  HL,DE    FFF8 11 03 08 32 FF FF 02 F8  0000  0000
2022  38FD    JR   C,FF     FFF8 11 03 08 32 FF FF 02 F8  0000  0000
2021  19      ADD  HL,DE    FFF8 11 03 08 32 FF FF 02 F7  0000  0000
2022  38FD    JR   C,FF     FFF8 11 03 08 32 FF FF 02 F7  0000  0000
2021  19      ADD  HL,DE    FFF8 11 03 08 32 FF FF 02 F6  0000  0000
2022  38FD    JR   C,FF     FFF8 11 03 08 32 FF FF 02 F6  0000  0000
2021  19      ADD  HL,DE    FFF8 11 03 08 32 FF FF 02 F5  0000  0000
2022  38FD    JR   C,FF     FFF8 11 03 08 32 FF FF 02 F5  0000  0000
2021  19      ADD  HL,DE    FFF8 11 03 08 32 FF FF 02 F4  0000  0000
2022  39FD    JR   C,FF     FFF8 11 03 08 32 FF FF 02 F4  0000  0000
2021  19      ADD  HL,DE    FFF8 11 03 08 32 FF FF 02 F3  0000  0000
2022  38FD    JR   C,FF     FFF8 11 03 08 32 FF FF 02 F3  0000  0000
2021  19      ADD  HL,DE    FFF8 11 03 08 32 FF FF 02 F2  0000  0000
2022  38FD    JR   C,FF     FFF8 11 03 08 32 FF FF 02 F2  0000  0000
2021  19      ADD  HL,DE    FFF8 11 03 08 32 FF FF 02 F1  0000  0000
2022  38FD    JR   C,FF     FFF8 11 03 08 32 FF FF 02 F1  0000  0000
2021  19      ADD  HL,DE    FFF8 11 03 08 32 FF FF 02 F0  0000  0000
2022  38FD    JR   C,FF     FFF8 11 03 08 32 FF FF 02 F0  0000  0000
2021  19      ADD  HL,DE    FFF8 11 03 08 32 FF FF 02 EF  0000  0000
```

Another type of library is a *macro library*. This contains not subroutines but sections of programs which are fitted into a program, as required, to perform a specific function. It is not intended here to examine the advantages of macros over subroutines or vice versa.

*Linking*

The linker program is a valuable tool in the production of structured programs. It allows the convergence of a number of small programs coupled with a library into the main object code. So a large program can be written and tested as a series of smaller ones. The linker will then join these together, allocating memory addresses as it goes, introducing the appropriate library routines into the correct places in the program and putting the complete object code into the specified file on the user disc.

A list file is produced giving the assigned addresses of the symbols used in the program.

The source and listing files should be arranged to have an origin of 0000H for these operations.

### Using the development system for program testing

In the 8550 development system there are three levels of operation for the testing of programs.

First, the program can be tested in the development system under the control of a *debug* program.

Second, the program can be run in the development system but by means of an in-circuit emulation (ICE) probe the program operates on the 'target hardware'. This allows the final product hardware – I/O etc. – to be tested under control of the debug program. The ICE probe replaces the central processing unit (CPU) in the target system but the program is still stored in the development system.

The third mode of operation is for the program to reside in the target hardware, but the CPU is still under control of the debug program; that is, all the hardware except the CPU is being used.

In this way the program and hardware can be tested under complete control. However, it should be noted that if critical timing routines are incorporated in the program these may not execute correctly, as the emulator processor may have to run more slowly owing to the extra length of connections to the processor and the operation of the debug program.

### Exercises

It is essential that you become familiar with the facilities offered by your particular system, as you will need to use it a great deal in the following chapters.

As different readers may have access to different systems it is not appropriate in this book to provide detailed instructions on which keys to press and in which order. This information is available in the manual supplied with your system.

The following set of exercises were developed for use with HEKTOR but could be implemented on any system, given possibly minor modifications.

**Exercise 1.1**

1   Connect up the system as described in the manual, switch on the power, and initialize the system (on HEKTOR, press RESET).

2   Set the system into EDIT mode (on HEKTOR press E, space and then RETURN).

3   Set the editor to receive program lines (on HEKTOR press I, space and then RETURN).

4   Type in the simple program in Figure 1.2 line by line, pressing RETURN at the end of each line.

5   Whilst typing in one line, delete some characters and retype them to demonstrate how errors could be corrected. (On HEKTOR use the ⬅ and then retype.)

6   When all the lines have been entered, quit the editor mode (on HEKTOR press BREAK).

7   List the program on the screen (on HEKTOR press Pl, 10, RETURN. As there are ten program lines the display should now look like Figure 1.2).

8   If an error has been made in any line, it will be necessary to correct that line. If the error is in line 3, instruct the system that you wish to edit that line (on HEKTOR press E3, RETURN).

9   Change the offending character (on HEKTOR, as in step 5).

10   Now add two new lines to the program:
   (a)   Add a new line above the current program
      ORG 3100H (see Section 1.3).
   (b)   Add a new line below the program
      END (see Section 1.3).
   (On HEKTOR, call the editor; press E, space, RETURN. Specify you wish to insert lines; press I, space, 0, space, RETURN (this specifies that you wish to insert a new line 0). Then type the new line followed by RETURN and BREAK to quit the editor. The second new line will be line 12 and this can be added in the same way, except that you would specify the new line by pressing I, space, 12, RETURN.) The program will now be as in Figure 1.3.

11   Save the source program you have now entered on to tape or disc – the user manual will explain how this is done.

12   Clear the computer and load the source program from the tape, check that it lists correctly (Figure 1.4).

13   Assemble the source program (on HEKTOR press A, RETURN or, if you also wish a listing file, A L, RETURN).

14   Save the assembled object program on tape, clear the computer and then reload the object program (on HEKTOR use the A T command).

15   Run the assembled program (on HEKTOR press A M to assemble and return command to the system monitor, and then press G for GO).

*Exercise 1.2*

When a program is written in assembler language, the actual memory address where the code or data is to reside is usually unknown. A *symbol* is used to replace the address. This symbol is assigned an appropriate address by the assembler program. In our example program there are already three symbols – START, TOTAL and STOP. However, to make the program more readable and possibly portable between machines, certain addresses can be declared as constants at the beginning of the program. Then, if the program needs to be run on another machine, it is only necessary to change a few statements and then reassemble. This saves searching the whole program for particular addresses.

Consider the revised program:

```
STACK:      EQU 3FFFH
DATA:       EQU 3800H
MONITOR:    EQU   57H
;
            ORG 3100H
START:      LXI SP, STACK; set up user stack
            LXI H, DATA
TOTAL:      MOV A, M
            ORA A
            INX H
            ADC M
            RAR
            INX H
            MOV M, A
STOP:       JMP MONITOR
            END
```

The use of EQU ('takes the value of' or 'equates to') tells the assembler that every time it finds this symbol it should replace it with that address or value.

EQU values may be any length up to 16 bits, but lesser values will assemble to 16-bit numbers with leading zeros. For example, 57H will assemble to 0057H but this will fit an 8-bit register, if necessary, as the leading zeros are ignored.

Use the editor to modify your program then re-assemble it. The modified program and the new listing should appear as in Figure 1.9

*Exercise 1.3   Data definition*

If a program calls for a list of constants or a look-up table, then the DB (define byte) of DW (define word) is used. These assembler directives allow the definition of data in either 8-bit or 16-bit form to be stored in a program.

Figure 1.9   *A program using symbolic addressing*

```
0001 STACK:EQU 3FFFH ;STACK TOP
0002 DATA: EQU 3800H ; DATA BUFFER START
0003 MONI: EQU 57H ;MONITOR WARM START
0004 ;
0005 ORG 3100H
0006 ;
0007 START: LXI SP,STACK ;SET UP STACK
0008  LXI H,DATA ;SET DATA START ADDR.
0009 TOTAL: MOV A,M ;GET FIRST NUMBER
0010  ORA A ;CLEAR CARRY
0011  INX H ;POINT TO NEXT NO.
0012  ADC , ;ADD NUMBER
0013  RAR ;DIVIDE BY 2
0014  INX H ; MOVE POINTER
0015  MOV A,M ;SAVE ANSWER
0016 STOP: JMP MONI ;END OF PROGRAM
0017  END
```

```
                    PASS 1

                    PASS 2

                    PASS 3
3100        0001 STACK:   EQU  3FFFH      ;STACK TOP
3100        0002 DATA:    EQU  3800H      ;DATA BUFFER START
3100        0003 MONI:    EQU  57H        ;MONITOR WARM START
            0004 ;
3100        0005          ORG  3100H
            0006
3100 31FF3F 0007 START:   LXI  SP,STACK   ;SET UP STACK
3103 210038 0008          LXI  H,DATA     ;SET DATA START ADDR.
3106 7E     0009 TOTAL:   MOV  A,M        ;GET FIRST NUMBER
3107 B7     0010          ORA  A          ;CLEAR CARRY
3108 23     0011          INX  H          ;POINT TO NEXT NO.
3109 8E     0012          ADC  M          ;ADD NUMBER
310A 1F     0013          RAR             ;DIVIDE BY 2
310B 23     0014          INX  H          ; MOVE POINTER
310C 7E     0015          MOV  A,M        ;SAVE ANSWER
310D C35700 0016 STOP:    JMP  MONI       ;END OF PROGRAM
3110        0017          END
ORIGIN: ENTRY = 3100; 3100
STACK = 3FFF   DATA = 3800   MONI = 0057   START = 3100
TOTAL = 3106   STOP = 310D
```

The DB directive stores the specified data in consecutive memory locations starting with the current setting of the program counter.

*Example*

(a)   Optional label   DB   byte, (byte, byte ......)

(b)   LUKTB:           DB   30H, 31H, 32H, 33H, 34H,
                            35H, 36H, 37H, 39H, 41H,
                            42H, 43H, 44H, 45H, 46H.

The DB may also be used to define the data as strings of ASCII characters. These will be assembled as their hex values.

*Example*

(c)   STRN:           DB   'OPERATION';

Figure 1.10   *A program showing the use of data definition directives*

```
0001 ; TITLE - PUTMSG.
0002 ;ROUTINE TO PRINT A MESSAGE
0003 ;ON HEKTOR'S SCREEN
0004 ;THE MESSAGE MAY BE CHANGED BY
0005 ;CHANGING THE TEXT AT MSG1 AND MSG2
0006 ;LABLES. TAKE CARE NOT TO MAKE
0007 ;THE LINES TOO LONG OR THE
0008 ;ASSEMBLER WILL CUT THEM OFF
0009 ;OR TRUNCATE THEM
0010 ;
0011 PRMES: EQU 30AH ;MONITOR CALL
0012 PRNL: EQU 2DAH ;---DITTO---
0013 MONW: EQU 57H ;MONITOR WARM START
0014 ;
0015 DRG 3800H
0016 ;
0017  CALL PRNL ;START NEW LINE
0018 PUTM: LXI  H,MSG1 ;POINTER TO 1ST. LINE
0019  CALL PRMES ;WRITE LINE OF TEXT
0020  CALL PRNL ;NEW LINE
0021  LXI H, MSG2 ;POINTER TO NEXT LINE
0022  CALL PRMES ;AND PRINT IT
0023  CALL PRNL ;ANOTHER NEW LINE
0024  JMP MONW ;AND RETURN TO MON.
0025 ;
0026 ;TEXT FOR THE MESSAGE--
0027 ;
0028 MSG1: DB 'THIS ROUTINE SENDS A MESSAGE'
0029  DB 0 ;TERMINATOR FOR PRMES ROUTINE
0030 MSG2: DB  'TO THE USER'
0031  DB 0 ;AS ABOVE
0032 ;◆◆◆◆ NOTE THE MESSAGE MUST BE ENCLOSED ◆◆◆
0033 ; IN SINGLE QUOTATION MARKS (SHIFT 7) AND
0034 ; NOT DOUBLE QUOTES (SHIFT 2).
0035 END
```

Figure 1.11   *A multibyte addition program (source:* Intel 8085 Language Manual)

```
0001 ;MULTI-BYTE ADDITION PROGRAM
0002 ;
0003 ;NUMBERS ARE STORED IN BUFFERS -
0004 ;FIRST AND SECND. RESULT IS STORED
0005 ; IN FIRST.
0006 ;
0007  ORG 3800H
0008 ;
0009 MADD: LXI B,FIRST ;B & C ADDRESS FIRST
0010  LXI H,SECND ;H & L SECOND ADDRESS
0011  XRA A
0012 LOOP: LDAX B ;LOAD BYTE OF FIRST
0013  ADC M ;ADD BYTE OF SECOND
0014 ;       WITH CARRY
0015  STAX B ;STORE RESULT AT FIRST
0016  DCR E ;DONE IF E=0
0017  JZ DONE
0018  INX B ;POINT TO NEXT BYTE
0019  INX H ;
0020  JMP LOOP
0021 DONE: JMP 0057H ;RETURN TO MONITOR
0022 ;
0023 FIRST: DB 90H,0BAH,84H
0024 SECND: DB 8AH,0AFH,32H
0025 END
```

This will assemble as:

4F    STRN:              DB        'OPERATION';

The program shown in Figure 1.10 writes a message on HEKTOR's screen. Enter, assemble and run this program.

This second example shows the use of the DB directive. It is a multibyte addition program (see Figure 1.11).

### Exercise 1.4   *Providing storage*

Some programs require an amount of RAM to be provided for the temporary storage of variables. For example, this may be used as a keyboard buffer when a string of characters are required to be input.

Memory may be required for use as a 'notebook' or 'scratch pad' for storage of partial results during a calculation.

The provision of this area of RAM is defined by the DS – define storage assembler directive. This allocates a consecutive area of memory from the current contents of the location counter to the current counter, plus the operand of the DS directive.

*Example*
2100      K BUFF       DS       20H;      keyboard buffer.

This would provide 32 (20H) bytes of storage and the location counter would be advanced to 2121H for the next instruction or directive.

### Exercise 1.5

The program shown in Figure 1.12 receives a series of inputs from the HEKTOR keyboard and stores them in a buffer until a ⌜RETURN⌝ is received. It then echoes the whole string back on the visual display unit (VDU).

### Exercise 1.6

Rewrite the program in Exercise 1.5 to convert the ASCII 0 to 9 and A to F into their binary equivalents stored in the buffer.

### Exercise 1.7   *Practical debugging*

In the programs which follow (Figures 1.13, 1.14 and 1.15), both syntax and logical errors occur. Use the HEKTOR facilities to correct the errors and to test the programs.

A description of the purpose of each program is included in the preamble of each.

Figure 1.12   *Example program ECHO*

```
0001 ;TITLE - ECHO
0002 ; WRITTEN MARCH 1982 - P.D.S.
0003 ;
0004 ;          FUNCTION
0005 ;THIS PROGRAM TAKES A LINE OF TEXT FROM
0006 ;THE HEKTOR KEYBOARD, SAVES IT IN A
0007 ;BUFFER, AND ON RECEIPT OF A CARR. RETURN
0008 ;PRINTS THE CONTENTS OF THE BUFFER
0009 ;ON THE V.D.U.
0010 ;
0011 ;INPUT :-NOTHING
0012 ;OUTPUT :-NOTHING
0013 ;
0014 ;USES REGISTERS :-A,F,B,HL
0015 ;NOTE ...SETS STACK POINTER
0016 ;
0017 ORG 3800H ;START OF PROGRAM
0018 ;
0019 ;TABLE OF EQUATES AND CONSTANTS
0020 ;==============================
0021 ;
0022 ;(MONITOR CALLS)
0023 MONW: EQU 57H ;RE-ENTRY POINT
0024 PRNL: EQU 2DAH ;CRLF
0025 PRMES: EQU 30AH ;PRINT STRING
0026 KR: EQU 5BEH ;KEYBOARD HANDLER
0027 KLUC: EQU 660H ;LC-UCV CONVERTER
0028 TV: EQU 6C0H ;VDU HANDLER
0029 ;
0030 CR: EQU 0DH
0031 EOT: EQU 00 ;END OF TEXT MARKER
0032 ;
0033 ;START OF PROGRAM
0034 ;
0035 ECHO: LXI SP,3FF0H ;SET STACK POINTER
0036   CALL PRNL ;NEW LINE
0037   MVI B,20H ;CHAR. COUNTER
0038 GETEM: LXI H,KBUF ;POINTER TO BUFFER
0039 GETEM2: CALL KR ;GET CHARACTER
0040   CPI CR ;END OF INPUT
0041   JZ SCRN ;YES, SO ECHO
0042   CALL KLUC ;CLEAR LC LETTERS
0043   CALL TV ;ECHO KEY TO SCREEN
0044   MOV M,A ;SAVE IN BUFFER
0045   INX H ;INC POINTER
0046   DCR B ;DEC COUNTER
0047   JNZ GETEM2 ;AND CONTINUE
0048 SCRN: MVI A,EOT ;ADD EOT MARK
0049   MOV M,A ;
0050 LXI H,KBUF ;AND PRINT THE
0051   CALL PRNL ;CONTENTS OF
0052   CALL PRMES ;THE BUFFER
0053   JMP MONW ;AND RETURN TO MON.
0054 ;
0055 ;PROVIDE SPACE FOR THE BUFFER
0056 ;
0057 KBUF: DS 20H ;BUFFER FOR 32 BYTES
0058   END
```

Figure 1.13    *Faulty program*
*number 1: FLASHER*

```
0001 ; TITLE FLASHER
0002 ;
0003 ;VERSION 1.0
0004 ;
0005 ;PROGRAM FOR DEBUGGING
0006 ;
0007 ;FUNCTION:- THIS PROGRAM FLASHES L0 ON
0008 ;THE PERIPHERIAL BOARD
0009 ;
0010 ;THIS PROGRAM CONTAINS THREE >BUGS<
0011 ;WHICH MUST BE CLEARED FOR PROPER
0012 ;OPERATION
0013 ;
0014 ;CONNECT THE BOX TO HEKTOR BEFORE SWITCHING
0015 ;THE COMPUTER ON
0016 ;
0017  ORG 3800
0018 ;
0019 MONITOR CALLS
0020 MONW:EQU 57H
0021 SDEL:EQU 732H
0022 ;
0023 I/O ADDRESSES
0024 CON:EQU 40H ;PORT CONTROL ADDRESS
0025 DTA:EQU 42H
0026 ;
0027 START:LXI SP,3000H ;SET SP
0028  MVI A,0FH ;0F=CONTROL WORD FOR PORT
0029  OUT CON ,
0030 ON: MVI A,1 ;SWITCH FOR LED0
0031  OUT DTA ;SWITCH IT ON
0032  CALL SDEL ;WAIT
0033  XRA A ;CLEAR SWITCH
0034  OUT DTA ;TURN LED OFF
0035  CALL SDEL ;WAIT
0036  JMP ON
0037  END
```

Figure 1.14    *Faulty program*
*number 2: HEXCON*

```
0001 ;TITLE HEXCON
0002 ;VERSION 1.0
0003 ; WRITTEN 9.81 - P.D.S.
0004 ;
0005 ;FUNCTION:-
0006 ; ASCII TO HEX (BINARY) CONVERSION ROUTINE
0007 ; MAY BE USED AS A SUB-ROUTINE.
0008 ; TAKES TWO ASCII CHARACTERS FROM
0009 ; THE KEYBOARD AND CONVERTS THEM INTO
0010 ; AN 8-BIT NUMBER IN REGISTER A,
0011 ; AND WRITES IT ON THE SCREEN
0012 ;
0013 ; INPUT NOTHING
0014 ; OUTPUT 8-BIT NUMBER IN A
0015 ;
0016 ; USES REGISTERS :- A,HL
0017 ;
0018  ORG 3800H ;START ADDRESS (MAY BE MOVED)
0019 ;
0020 ; TABLE OF EQUATES ETC.
0021 ; ======================
0022 ;
0023 MONW; EQU 57H
0024 OUTS: EQU 2E7H ;PRINT A SPACE
0025 NEWL:EQU 2DAH ;NEW LINE
0026 PRB: EQU 35DH ;PRINT REG. A
0027 KR: EQU 5BEH ;KEYBOARD INPUT
0028 KLUC: EQU 660H ;UC-LC CONVERTER
```

*continued*

```
0029 TV:EQU 6COH ;VDU OUTPUT
0030 ;
0031 HEXCON: CALL  NEWL ;NEW LINE
0032  MVI A,3FH ;ASCII ?
0033  CALL TV
0034  CALL OUTS ,OUTPUT A SPACE
0035  CALL KR ;GET ASCII
0036  CALL CONV
0037  RLC ;POSITION UPPER NIBBLE
0038  RLC ;
0039  RLC ;
0040  RLC ;
0041  ANI FOH ;CLEAR LOWER NIBBLE
0042  MOV B,A ;AND SAVE IT
0043  CALL KR ;GET NEXT
0044  CALL CONV ;CONVERT
0045  ANI OFH ;CLEAR RUBBISH
0046  ADD B ;FORM BYTE
0047  CALL NEWL ;CRLF
0048  CALL PRB ;PRINT NUMBER
0049  JMP MONW ;RETURN (CHANGE TO RETURN
0050 ;                   FOR SUBROUTINE USE)
0051 ;
0052 ;CONVERSION ASCII TO HEX.
0053 ;
0054 CONV:CALL KLUC ;CONVERT TO LOWER CASE
0055  SUI 30H ;SUB 48
0056  JM NOTX ; NOT HEX.
0057  CPI OAH ;NUMBER< 10?          (ED: NO CHEVRON)
0058  JC GOTTEM ;YES
0059  CPI 11H ;CHECK FOR 9-A
0060  JC NOTX ;ERROR, BETWEEN 9 & A
0061  CPI 17H ; NUMBER  >F?          (ED: NO CHEVRON)
0062  JNC NOTX ;YES
0063  SUI 7 ;CONVET AND
0064 GOTTEM: RET ;RETURN
0065 NOTX:JMP MONW ;NOT HEX
0066  END
```

Figure 1.15  *Faulty program number 3: BCDCON*

```
0001 ;TITLE BODOON
0002 ; VERSION 1.1
0003 ;
0004 ;PURPOSE:-
0005 ;  CONVETS THE NUMBER IN ACCUMULATOR
0006 ;  TO A 3-DIGIT BCD NUMBER STORED IN
0007 ;  THREE CONSECUTIVE MEMORY LOCATION
0008 ;  BCD2, BCD1 & BCD0
0009 ;
0010 ; INPUT - NUMBER FOR CONVERSION
0011 ;          IN THE ACCUMULATOR
0012 ; OUTPUT - CONVERTED NUMBER IN MEMORY
0013 ;
0014 ; USES B & HL
0015 ; ACC. IS PRESERVED.
0016 ;
0017 ;CONSTANTS :-
0018 HUN: EQU 100
0019 TEN:EQU 10
0020 ;
0021 ;SPACE FOR ANSWER :-
0022 BCD0:EQU 3A00H
0023 BCD1:EQU 3A01H
0024 BCD2:EQU 3A02H
0025 ;
0026 ;MONITOR CALL
```
*continued*

```
0027 MONW:EQU 57H
0028  ORG 3800H
0029 ;
0030 BCDCON: PUSH PSW ;SAVE NUMBER
0031  PUSH PSW ;TWICE.
0032 ;
0033 ;INITIALIZE THE ANSWER
0034 ;
0035 CLEAR:LXI H,BCDO ;LSD
0036  XRA A ;CLEAR A
0037  MVI B,3 ;NUMBER OF LOCATIONS
0038  MOV M,A ;CLEAR LOCATION
0039  INX H ;AND ON TO THE NEXT
0040  DCR B ;
0041  JNZ CLEAR ;
0042  DCX H ;ADJUST HL
0043  POP PSW ;ALL DONE RESTORE NO.
0044 MVI B,HUN ;B=100
0045 ANO1:SUB B ;A=A-100
0046  JM NHUN ;A IS< 100
0047  INR M ;ANS = ANS + 100
0048  JMP ANO1 ;TRY AGAIN
0049 NHUN:ADD B ;RESTORE A
0050  DCX H ;POINT TO TENS
0051  MVI B,TEN ,
0052 ANO2:SUB B ;B=B-10 ETC.
0053  JM NOTEN ;
0054  INR M ;
0055  JMP ANO2 ,
0056 NTEN:ADD B ;TENS FINISHED SO
0057  DCX H ;SAVE REMAINDER IN UNITS
0058  MOV M,A ;
0059  POP PSW ;RESTORE A
0060  JMP MONW ; OR RETURN
0061 ;
0062  END
```

## 1.5  Programming techniques

When constructing programs or routines it is very tempting just to write the source text, assemble it and use the object program. However, it is very important that any software is fully documented so that it is understandable by others and also to the programmer when returning to a routine some months later. The use of comment lines permit explanation of a program; in fact a line of program should never be written unless a comment is added. Further, each program should be preceded by an introduction explaining what it does, which registers it uses, which are destroyed and any other routines it calls. Compare the programs in Figure 1.16 and Figure 1.17; which is the more explanatory?

Note the main sections of the preamble to program:

*Title*
This is the name of the program which appears in the heading of each page of the program listing.

Figure 1.16 *An undocumented program*

```
0001 COP: MOV A,M
0002    STAX D
0003    DCX B
0004    INX H
0005    INX D
0006    MOV A,B
0007    ORA C
0008    JNZ COP
0009    RET
```

Figure 1.17 *A documented program (LDIR/copy)*

```
0001 ; NAME = COPY
0002 ;VERSION 1.0
0003 ;
0004 ;THIS ROUTINE DUPLICATES THE LDIR
0005 ;INSTRUCTION OF THE ZILOG Z80 ON THE
0006 ;INTEL 8080/8085 PROCESSORS
0007 ;
0008 ;WRITTEN...JAN 1980 BY FRED BLOGGS.
0009 ;
0010 ;INPUT.. ENTER WITH :-
0011 ;        DATA SOURCE ADDRESS IN HL
0012 ;        DATA DESTINATION ADDRESS IN BE
0013 ;        AND THE NUMBER OF BYTES
0014 ;        TO BE MOVED IN BC.
0015 ;
0016 ;OUTPUT..NOTHING
0017 ;
0018 ;DESTROYS ALL REGISTERS..
0019 ;
0020 COP: MOV A,M ;GET BYTE IN ACC.
0021    STAX D ;STORE BYTE IN DEST.
0022    DCX B ;DECREMET BYTE COUNTER
0023    INX H ;MOVE POINTERS.
0024    INX D ;
0025    MOV A,B ;TEST FOR LAST BYTE
0026    ORA C ;WHEN B=C=0
0027    JNZ COP ;NOT LAST BYTE
0028    RET ;WAS LAST BYTE
0029    END
```

### Date
This should be the date of the creation of the program. Progress of the program can be checked against the version number and date.

### Version number and revision number
This indicates the status of the program. Version 1.0 is the original unmodified program. Any changes to sections or subroutines will cause the revision number to increase – i.e. V1.1, V1.2, etc. Major changes in program structure will warrant a new version number – e.g. V2.0. The date of revision or change of version should also be shown.

### Programmer name(s)
In the case of large programs the job will be split between a number of programmers. The inclusion of the name indicates responsibility.

*Description*

Explains briefly the function of the program and its relationship (if any) with its parent program.

*Input*

This section states in which registers or memory locations the routine expects to find its parameters for correct execution. The calling program must load these stores before calling the routine. In this example the HL register pair holds the source address of the data to be moved, the DE register pair holds the destination address for the data, and register B holds the number of bytes to be moved.

*Output*

This lists the status of registers and any modified memory locations at the end of the routine. Any register contents detroyed as a result of running the routine must be saved before it is executed.

## 1.6   Conclusion

This chapter has introduced the use of development systems as an aid to writing, testing and updating programs and the testing of hardware under controlled conditions.

Two types of system have been examined – HEKTOR, a simple system for program development and test, and the Tektronix 8002 microprocessor lab., which allows not only testing but also controlled operation of the target hardware.

The main units of a development system have been identified and systems have been used to write, debug and test simple programs.

If used confidently, a good development system will repay its cost many times over in saved time and effort, but knowledge of its facilities and operation is essential if it is to provide this.

# Chapter 2 Input/output techniques

*Objectives of this Chapter* *When you have completed this chapter, you should be able to:*

1 *Identify I/O devices.*
2 *Test software with appropriate hardware peripherals.*
3 *Use a parallel port to interface to external devices, such as keyboards and motors.*
4 *Select hardware to drive a stepper motor.*
5 *Use a serial port to communicate with another computer.*

## 2.1 Introduction

When microprocessors are used to control systems, some method of interface between the computer and peripheral is required. There may be time-related problems to overcome owing to the difference in the speed of operation of the controlled device and the computer, and there will almost certainly be a power level interfacing problem to be solved.

Most small computer systems allow for the connection of other devices. Depending on the machine, this is usually done by providing a socket or port (or number of ports), typically either 8 or 16 bits wide.

Each pin in each port may be placed at logic 1 or 0 by specifying the appropriate binary word that is to appear at the port. These words can be changed under computer control and so an external device connected to the port can receive a number of signals which can control its operation.

Similarly, signals passed from a device to the port can be read by the computer and analysed to determine future behaviour. In practice, input/output (I/O) buffers are used to raise the current that may be drawn from the port, protect the computer and remember the word written to the buffer until this is superseded by another word.

In a number of systems, the ports are memory mapped, that is, they each have a memory address allocated to them. If you move a particular word to one of their addresses, rather than storing that word in RAM it is placed on the relevant port. Reading the port can be accomplished simply by accessing the appropriate address.

However, the 8085 processor used in HEKTOR computer has separate input and output instructions for accessing the ports.

## 2.2   An interfacing device in the HEKTOR peripheral unit

The HEKTOR system has some interfacing devices built into its peripheral box. One of these is the 8155 RAM I/O timer device (let's call it a RIOT!) The full data sheet for this can be found in Appendix 2.

The RIOT has three ports two 8 bits wide (port A and port B) and one 16 bits wide (port C). Ports A and B are programmable to be either input or output; port C can be programmed to be either input, output or a control port.

### *Programming the port for output*

Figure 2.1 *Bit assignment of 8155 command register*

To program the port function, the correct *control word* must be written to the command register. This is an 8-bit register and the assignment of each bit is shown in Figure 2.1.

The bits marked X are 'don't care' for the moment and will be set to 0.

The programming for port C provides the four alternatives given in Table 2.1.

**Table 2.1**   *8155 port C I/O programming*

| D3 | D2 | Resultant configuration | |
|----|----|------|------|
| 0 | 0 | All pins inputs | |
| 0 | 1 | All pins outputs | |
| 1 | 0 | D0 | A interrupt |
|   |   | D1 | A buffer full |
|   |   | D2 | A strobe (active low) |
|   |   | D3–5 | output |
| 1 | 1 | D0–2 | as above) for port B |
|   |   | D3–5 | as above) |

The ports are accessed by the I/O instructions and have the address arrangements given in Table 2.2.

**Table 2.2**   *8155 I/O addresses*

| I/O address A7                A0 | Function |
|----|----|
| X  X  X  X  X  0  0  0 | Command register |
| X  X  X  X  X  0  0  1 | Port A |
| X  X  X  X  X  0  1  0 | Port B |
| X  X  X  X  X  0  1  1 | Port C (D0 to D5 only) |

The high-order address lines are decoded on HEKTOR's peripheral board to give the complete I/O addresses of 40H to 43H for this device.

### Simple control of a d.c motor

A simple control function could be to turn a 5 V d.c. motor on, let it run for a fixed period, and then switch it off. This could be accomplished by connecting the power supply for the motor to one of the buffered output pins of the computer port, once a word has been 'written' to that port setting the appropriate pin to logic 0. A word can then be written to the port setting that pin to logic 1, which will start the motor; then, after the required period, that pin can again be switched to logic 0, turning the motor off.

### Exercise 2.1

The HEKTOR kit is designed to allow this type of exercise to be carried out easily – the motor is controlled from bit 5 on port B (PB5) via its buffer amplifiers.

This means that port B is to be set for output; that is, D1 in the command register *must* be a 1. For the purpose of this exercise both the other ports will be set up as inputs, so that control word is 02H.

The program presented (Figure 2.2) runs the motor for 30 seconds and then switches it off.

Figure 2.2   *Program RUNMOTOR*

```
0001 ;TITLE RMOT
0002 ;
0003 ;ROUTINE TO RUN THE MOTOR
0004 ;ON THE HEKTOR PERIPHERIAL BOARD
0005 ;FOR A SET PERIOD.
0006 ;ENTER THE TIME * 4 IN LOCATION 3850H
0007 ;
0008 ;INPUT TIME IN LOCATION ''ONTIM''
0009 ;
0010 ;OUTPUT NOTHING
0011 ;
0012 ;REGISTERS USED - A, B, D, E, HL
0013 ;
0014 ;TABLE OF EQUATES AND CONSTANTS
0015 ;
0016 CNTRL:EQU 40H
0017 MPORT:EQU 42H
0018 ON:EQU 0D0H
0019 OFF:EQU 0F0H
0020 ;
0021 ONTIM:EQU 3850H
0022 ;
0023 ORG 3800H
0024 ;
0025 ;INITIALISE THE RIOT
0026 ;
0027 INIT:MVI A,2 ;CONTROL WORD
0028  OUT CNTRL ;SEND IT TO THE PORT
0029 ;
0030 ;TURN THE MOTOR ON
0031 ;
```

*continued*

```
0032 MVI A,ON ;SET PB5 + 0 FOR ON
0033  OUT  MPORT ;SWITCH THE MOTOR ON
0034  CALL DEL3 ;WAIT FOR ONTIME
0035  MVI A,OFF ;PB5=1 FOR OFF
0036 OUT MPORT ;SWITCH MOTOR OFF
0037  HLT ;FINISH
0038 ;
0039 ;SOFTWARE TIMER
0040 ;
0041 ;DEL3:LDA ONTIM ;GET TIME CONSTANT
0042 ;
0043 ;INNER LOOP OF 17 MACHINE CYCLES * LOOP
0044 ;CONSTANT OF B0C0 HEX.
0045 ;
0046 DEL2: LXI D,-1 ;DECREMETER
0047  LXI H,0B0C0H ;LOOP CONSTANT
0048 DEL1: DAD D ;DECREMENT HL
0049  JC DEL1 ;INNER LOOP
0050  DCR A ; DECREMT OUTER LOOP
0051  JNZ DEL2 ;AND CONTINUE IF NOT ZERO
0052  RET
0053  END
```

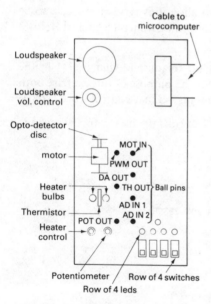

Figure 2.3 *Connecting the motor (source: Open University* HEKTOR User Manual*)*

To run the motor connect the PWM OUT pin to the MOTIN pin (see Figure 2.3).

Enter, assemble and load the program of Figure 2.2 and check its operation.

Try some different values of ONTIM.

Find the maximum and minimum times for the motor to run.

This system provides a simple ON/OFF control of the motor. If speed control is required then some method of changing the power supply to the motor must be arranged.

### Pulse width modulation control

A convenient way of controlling the speed of a d.c. motor is to vary the voltage powering it. This can be achieved by using a series of pulses, where the width of the pulses is under program control. The pulse time can be averaged to produce a d.c. level. If the pulses are narrow, and well spaced out, the average will be low and the motor will run slowly, whereas if the pulses form an almost continuous d.c. voltage the average will be 5 V and the motor will run at its maximum speed. This type of control is called pulse width modulation (PWM).

The voltage on PB5 must therefore be programmed to be a square wave where the pulse duration is set by the program (or perhaps calculated from the value of a separate input signal read from one of the other ports).

The hardware required to interface the PWM signal to the motor is very simple (Figure 2.4). An averaging network, comprising a few resistors and capacitors, smooths out the oscillations to their average value.

Figure 2.4 *PWM signals (source: Open University)*

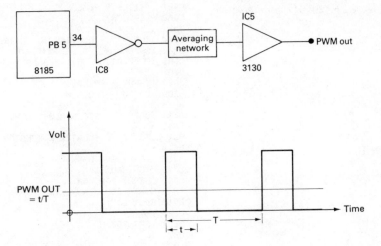

In order to ensure that the PWM signal oscillates between the well-defined power supply voltages (0 and 5 V), a buffer gate is interposed between the I/O pin of the 8155 and the averaging network. This gate is called an inverter, as its function is to supply a 0 value (0 V) when its input is 1 and a 1 value (5 V) when its input is 0. (The gate's symbol includes the small circle which indicates logical inversion of the binary signal.)

A second buffer follows the averaging network. Its function is to provide the current drive required by the motor subsystem. There is one peculiarity of the PWM circuit. Because of the inversion inherent in the buffer gate, the analogue output signal (PWM OUT) is a voltage proportional to the fraction of time that bit 5 of port B is 0, rather than 1. This poses no problem, provided the software writer is told!

### Controlling the motor speed

The speed of the motor is controlled by the proportion of time that bit 5 of port B is at logical zero (0 V). If the complete cycle of the output signal is considered divided into sixteen equal intervals (Figure 2.5)

Figure 2.5 *PWM timing (source: Open University)*

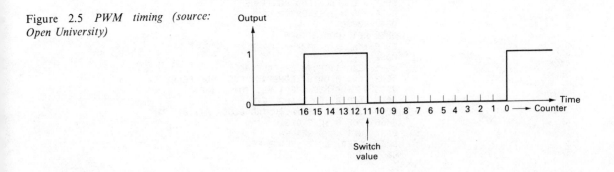

Figure 2.6   *PWM control program*

```
0001 ; TITLE = PWMCON
0002 ;VERSION 1.0/MAR 1982
0003 ;
0004 ;PURPOSE - ROUTINE TO CONTROL THE SPEED
0005 ;OF A SMALL D. C. MOTOR BY PULSE WIDTH
0006 ;MODULATION OF THE SUPPLY
0007 ;
0008 ;INPUT - SPEED IN LOCATION ''SPEED''
0009 ;OUTPUT - NOTHING
0010 ;
0011 ;USES REGISTERS A,B,C & HL
0012 ;
0013 ;TABLE OF EQUATES & CONSTANTS
0014 ;
0015 MTOFF: EQU 0F0H
0016 MTON:EQU 0D0H
0017 CNTRL: EQU 40H
0018 PORTB:EQU 42H
0019 BOUT:EQU 2
0020 ;
0021  ORG 3800H
0022 ;
0023 INIT:MVI A,BOUT ;SET UP PORT
0024 OUT CNTRL ;FOR OUTPUT
0025 LXI H,SPEED ;POINT TO MEMORY LOCATION
0026 ;
0027 ;CALCULATE OFF TIME
0028 ;
0029  MOV B,M
0030  MVI A,16 ;TOTAL CYCLE TIME
0031  SUB M ;OFF TIME =(16-ON TIME)
0032  MOV C,A ;SAVE ANSWER
0033 RUN: MVI A,MTON ;SWITCH THE MOTOR
0034  OUT PORTB ;ON
0035  CALL ONDEL ;WAIT FOR ON TIME
0036  MVI A,MTOFF ;TURN MOTOR
0037  OUT PROTB ;OFF
0038 CALL OFDEL ;WAIT FOR OFF TIME
0039  JMP RUN ;AND CYCLE AGAIN
0040 ;
0041 ;DELAY ROUTINES
0042 ;
0043 ONDEL:PUSH B ;SAVE TIME
0044  PUSH H ;SAVE POINTER
0045 ONDL:CALL DELAY1 ;MONITOR CALL
0046 DCR B
0047  JNZ ONDL ;AND CONTINUE UNTIL DONE
0048  POP H ;RESTORE
0049  POP B ;REGISTES
0050  RET ;AND RETURN
0051 OFDEL: PUSH B ;SAVE TIMES
0052  PUSH H
0053 OFDL:CALL DELAY1 ;WAIT ONE PERIOD
0054 DCR C ;DEC OFF TIMER
0055 JNZ OFDL ;AND CONTINUE
0056 POP H ;RESTORE REGS
0057 POP B ;
0058 RET ;AND RETURN
0059 ;
0060 ;
0061 DELAY1: PUSH B ;SAVE BC
0062  MVI B,0A7H ;CONSTANT FOR LOOP DELAY
0063 DEL1MS:NOP ;OF 1 MS. TOTAL INC.
0064  DCR B ;CALL AND RETURN
0065  JNZ DEL1MS ;ASSUMING CLOCK AT3.072 MHZ
0066  POP B ;RESTORE B
0067  RET ;AND RETURN
0068 SPEED: DS 1 ;SPACE FOR SPEED
0069 END
```

then the number of intervals that the output is zero will determine how fast the motor runs. This signal can be generated quite easily by a nested loop program. If the inner loop is set to 1 ms then control cycle time will be just over 15 ms.

Figure 2.6 illustrates a simple PWM control program. The speed is set by a value stored in variable location SPEED. The program uses this value as the ON time and calculates the OFF time for the waveform. The program then enters a continuous loop until reset.

### Exercise 2.2

Enter, assemble and test the program as given in Figure 2.6.

Try some different values of SPEED and determine maximum and minimum values.

Why is there a limit to how slowly the motor will run?

### Exercise 2.3

Modify the program to examine the binary inputs on S0 to S3 and use the input to control the speed of the motor. Note that:

1   As the port is 6 bits wide and only 4 bits are required, precautions must be taken to allow for data on the remaining two bits.
2   The control word for the port configuration B output, C input is 02H as before.

It is possible to use the switches S0 to S3 to provide the speed value rather than a variable written into the program. These switches are connected to bits 0 to 3 of port C.

To read the state of these switches it is first necessary to program port C for input. This is done by setting D2 and D3 to logic 0, so we write the control word 02 to address 40, then execute an IN 43H instruction. This puts the binary value of the switches into the A register. The number in the A register can then be stored in the location SPEED ready for access by the rest of the program.

## 2.3   Extending HEKTOR

The peripheral box supplied with the HEKTOR provides a simple way of controlling the devices mounted on the HEKTOR board. As for practical applications we shall require to interface HEKTOR with a wider range of peripherals, it is necessary to build a separate I/O unit. A number of suitable semiconductor chips are available and these can be assembled quite easily.

The simplest interface is an 8-bit latch such as the Intel 8212 (74S412) (see Appendix 2 for the data sheet). The 8212 is a unidirectional port

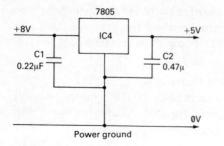

Figure 2.7 *Circuit of the 8212 peripheral board*

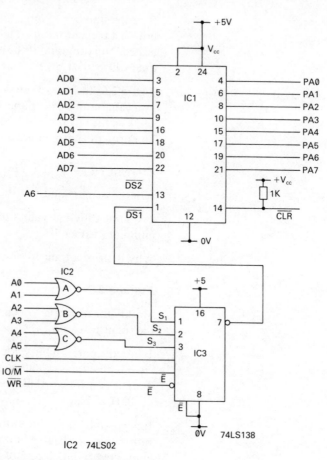

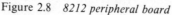

IC2  74LS02

Figure 2.8   *8212 peripheral board*

made up of eight latches with chip select and control logic. An even simpler I/O port could be constructed with, say 74LS75 I/Os, plus a few logic devices.

A circuit diagram of a possible new peripheral board is shown in Figure 2.7. Construction may be a wire wrap or on stripboard. The actual board layout may be designed by the student, but a suggested arrangement is shown in Figure 2.8.

### Circuit operation of new peripheral board

This description refers to Figure 2.8. The address/data bus is applied directly to the data input pins of the latches (pins AD0–AD7) but the information is not passed to the output pins until the strobe pins have been activated. There are two strobe pins – $\overline{DS2}$ and $\overline{DS1}$. $\overline{DS2}$ is fed by A6 and is required to be HIGH for strobing. $\overline{DS1}$ is fed with the decoded lower address bus. To allow a reasonable number of devices to be active on the I/O address bus, some decoding is necessary. In this instance A0 and A1, A2 and A3, and A4 and A5 are fed to two input NOR IC2A-C gates, 74LS02. The output of the NOR gate is HIGH only when both inputs are LOW, by selecting the $\overline{07}$ output of the 74LS138 three-to-eight line decoder, the required address is obtained, i.e. port address 40H.

Timing of the I/O strobe is provided by the WR and IO/M signals. These are fed to the enable lines on the 74LS138 I/O decoder; the third enable pin is connected to ground, as are the unused inputs on IC2.

When address 40H is accessed by an OUT instruction, data on the data bus will be latched on to the output lines of IC1. If required, the output pins may be reset by taking pin 14 (CLEAR) low by an external switch.

IC3 provides the necessary +5 V for the ICs.

A lead will need to be made up to connect the HEKTOR edge connector to the new I/O board via low-cost ten-way Molex connectors.

Table 2.3 lists the connections from the peripheral board to the processor board. Note that the low byte of the address bus is available (demultiplexed) on this connector. It must be remembered that these

**Table 2.3** *Microprocessor bus signals*

| Component side pin | 8085 pin | Signal name | Wiring side pin | 8085 pin | Signal name | |
|---|---|---|---|---|---|---|
| 1 | | 10 V unreg. d.c. power | 2 | | 10 V unreg. d.c. power | |
| 3 | | power ground | 4 | 20 | Signal ground | |
| 5 | 32 | $\overline{RD}$ (47 K pull-up) | 6 | 19 | AD7 | |
| 7 | 31 | $\overline{WR}$( 47 K pull-up) | 8 | 18 | AD6 | |
| 9 | 30 | ALE | 10 | 17 | AD5 | |
| 11 | 11 | $\overline{INTR}$ | 12 | 16 | AD4 | |
| 13 | 3 | RESETOUT | 14 | 15 | AD3 | |
| 15 | 37 | CLK | 16 | 14 | AD2 | |
| 17 | 28 | A15 | 18 | 13 | AD1 | |
| 19 | 39 | HOLD (3K9 pull-down) | 20 | 10 | INTR (3K9 pull-down) | |
| 21 | 38 | HLDA | 22 | 7 | RST7.5 (47 K pull-down) | |
| 23 | 35 | READY (3K9 pull-up) | 24 | 12 | ADD | |
| 25 | 34 | IO/$\overline{M}$ | 26 | 2) | | A0 |
| 27 | 27 | A14 | 28 | 5) | | A1 |
| 29 | 26 | A13 | 30 | 6) | pins on | A2 |
| 31 | 25 | A12 | 32 | 9) | IC 12 | A3 |
| 33 | 24 | A11 | 34 | 12) | (74LS373) | A4 |
| 35 | 23 | A10 | 36 | 15) | | A5 |
| 37 | 22 | A9 | 38 | 16) | | A6 |
| 39 | 21 | A8 | 40 | 19) | | A7 |

lines are not buffered in HEKTOR; hence only a very short length of ribbon cable can be used to connect the boards. A full parts list can be found in Chapter 6, together with a list of suppliers.

### Exercise 2.4　Using the simple port

This exercise shows the use of the simple port just described for on/off control. LEDs are used to indicate the state of the output pins, but with suitable buffering the controlled device could be almost anything from a lamp to a high-power motor.

Even LEDs need to be buffered, and the circuit of Figure 2.9 shows how this may be done. Breadboard techniques will provide a temporary arrangement, or stripboard could be used. One buffer and one LED is required for each output pin.

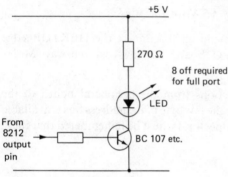

Figure 2.9　*LED output circuit*

```
0001 ;TITLE RUNL
0002 ;
0003 ;FUNCTION DEMONSTRATION  RUNNING LIGHT
0004 ;PROGRAM FOR SIMPLE I/O PORT
0005 ;
0006 PORT:EQU 40H
0007 ;
0008 ORG 3800H
0009 ;
0010 START:MVI A 1 ;SET BIT 0 TO 1
0011 PUT:OUT PORT ;TURN LAMP ON
0012 CALL WAIT ;KEEP IT ON FOR A SECOND OR SO
0013  RLC ;MOVE SIDEWAYS ONE POSITION
0014  JMP PUT
0015 WAIT :LXI D,-1;LOOP DECREMETER
0016  LXI  H,-1 ;DELAY TIME
0017 DECR:DAD D ;HL<HL-1   (ED NO CHEVRON)
0018  JC DECR ;KEEP LOOPING
0019  RET; WHEN DONE
0020  END
```

PASS: 1

Figure 2.10　*A simple running light program*

Figure 2.10 illustrates a simple 'running light' program for this board.

Enter, assemble and run this program then modify it for different speeds, direction etc.

Rewrite the program to make the light run alternately left and right.

Try different patterns in the A register and see that they appear correctly on the LEDs.

The simple port can be used for the control of motors but, as noted previously, power interfacing is also required. Figure 2.11 shows the circuit diagram of a suitable interface for the control of a motor. Two bits are used: the conditions produced are summarized in Table 2.4.

Figure 2.11   *A d.c. motor interface*

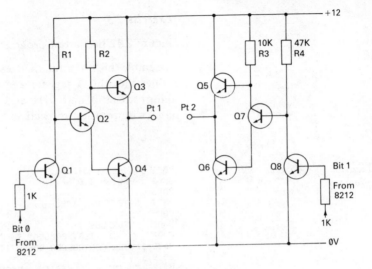

**Table 2.4**   *Motor control bit assignments*

| Bit | | Result |
|---|---|---|
| 0 | 1 | |
| 0 | 0 | Stationary |
| 0 | 1 | Anti-clockwise |
| 1 | 0 | Clockwise |
| 1 | 1 | Stationary |

### Direction control of a small d.c. motor

The specification for the motor used in this example is given in Appendix 2. The 12 V version is used here.

The direction control of a d.c. motor is fairly simple, in as much as to reverse the motor the polarity of the supply only needs to be reversed.

Additionally, as noted above, there is a power level change to be implemented.

*Circuit operation*
This description refers to Figure 2.11. When both inputs are at 0 or 1, then the potential on both motor terminals is the same and hence there is no rotation. If PA0=1 and PA1=0, then Q1 is turned ON removing the bias from Q2, so allowing Q3 ON and Q4 OFF. This allows port 1 to approach +12 V, whilst PA1=0 Q8 is OFF so keeping Q7; hence Q6 ON and Q5 OFF, so connecting port 2 to 0 V.

When the conditions on PA0 and PA1 are reversed, then the supply to the motor is effectively reversed.

### Exercise 2.5

Figure 2.12 lists a test program for the motor.

Assemble this program, then test it using the monitor. Enter different values into the A register and ensure that 00H and 03H cause the motor to stop and 01H and 02H cause the anti-clockwise and clockwise rotation respectively.

Figure 2.12   *A motor control program*

```
0001 ;TITLE :-MOTCON1
0002 ;VERSION 1.0
0003 ;
0004 ;DATE APRIL 1982
0005 ;
0006 ;  PURPOSE :-
0007 ;        TEST PROGRAM TO CHECK OPERATION OF D. C.
0008 ;MOTOR CONTROLLER OPERATION
0009 ;
0010 ;TYPING R CAUSES THE MOTOR TO ROTATE CLOCKWISE
0011 ;        L CAUSES THE MOTOR TO ROTATE ANTICLOCKWISE
0012 ;        S CAUSES THE MOTOR TO STOP
0013 ;
0014 MONW:EQU 57H
0015 NEWL:EQU 2DAH
0016 KB:EQU 5BEH
0017 KLUC:EQU 660H
0018 VDU:EQU 6C0H
0019 PORT:EQU 40H
0020 ;
0021  ORG 3800H
0022 ;
0023 INKEYS: CALL NEWL
0024  CALL KB ;GET KEY
0025  CALL KLUC;CONVERT TO LOWER CASE
0026  CALL VDU ;SHOW IT ON THE SCREEN
0027 ;
0028 ;WHAT WAS IT?
0029 ;
0030 CPI 52H ;AN R?
0031 JZ RIGHT;
0032 CPI 4CH ;AN L MAYBE?
0033 JZ LEFT ;
0034 CPI 53H ;STOP?
0035 JZ STOP
0036 JMP   INKEYS ;MUST HAVE BEEN RUBBUSH!
0037 RIGHT:MVI A,1 ;PORT = 0000 0001 BINARY
0038 SWITCH: OUT PORT ;SWITCH THE MOTOR ON
0039  JMP INKEYS: WAIT FOR NEXT COMMAND
0040 LEFT:MVI A,2; PORT = 0000  0010 BINARY
0041  JMP SWITCH
0042 STOP: XRA A;PORT TO = 0000 0000 BINARY
0043  JMP SWITCH
0044 END
```

### Exercise 2.6

Write a program to alternately rotate the motor for 30 seconds in each direction.

The 8212 peripheral board described provides a dedicated I/O port that needs no programming. However, it lacks the flexibility of the more complex devices, but there are some instances where this approach can be more cost effective.

## 2.4  Combined input/output devices

Another 8212 could be used to provide input to the processor, but a more versatile I/O unit can be made using a larger-scale chip such as the 8255A. This is a programmable I/O (programmable peripheral interface, PPI) device. It has three ports – A, B and C – similar to the 8155, but port C is now a full 8 bits wide.

### Ports A, B and C

The 8255A contains three 8-bit ports (A, B and C). All can be configured in a wide variety of functional characteristics by the system software, but each has its own special features or 'personality' to further enhance the power and flexibility of the 8255A.

#### Port A
One 8-bit data output latch/buffer or one 8-bit data input latch.

#### Port B
One 8-bit data input/output latch/buffer or one 8-bit data input buffer.

#### Port C
One 8-bit data output latch/buffer or one 8-bit data input buffer (no latch for input). This port can be divided into two 4-bit ports under the mode control. Each 4-bit port contains a 4-bit latch and it can be used for the control signal outputs and status signal inputs in conjunction with ports A and B.

These ports can be programmed as in the 8155 to provide a number of modes of operation.

### Mode selection

There are three basic modes of operation that can be selected by the system software.

Mode 0 – basic input/output
Mode 1 – strobed input/output
Mode 2 – bidirectional bus

When the reset input goes 'high' all ports will be set to the input mode (i.e. all twenty-four lines will be in the high impedance state). After the reset is removed the 8255A can remain in the input mode with no additional initialization required. During the execution of the system program any of the other modes may be selected using a single output instruction. This allows a single 8255A to service a variety of peripheral devices with a simple software maintenance routine.

The modes for port A and port B can be separately defined, while port C is divided into two portions as required by the port A and port B definitions. All of the output registers, including the status flip-flops, will be reset whenever the mode is changed. Modes may be combined so that their functional definition can be 'tailored' to almost any I/O structure. For instance, group B can be programmed in mode 0 to monitor simple switch closings or display computational results. Group A could be programmed in mode 1 to monitor a keyboard or tape reader on an interrupt-driven basis.

The mode is set by writing the appropriate CONTROL word to the command register.

The command word for mode 0 is identified by having a bit D7 set to 1, and the lower nibble defines the mode as summarized in Table 2.5.

**Table 2.5   Mode control summary**

0 = output
1 = input

| D7 | D6 | D5 | D4 | D3 | D2 | D1 | D0 |
|---|---|---|---|---|---|---|---|
| *Mode set bit* | *Mode no. upper group* | | *Port A* | *Port C D4–7* | *Mode no. lower group* | *Port B* | *Port D0–3* |
| 1 | 0 | 0 | 0 | 0 | 0 | 0 | 0 |
| 1 | 0 | 0 | 0 | 0 | 0 | 0 | 1 |
| 1 | 0 | 0 | 0 | 0 | 0 | 1 | 0 |
| 1 | 0 | 0 | 0 | 0 | 0 | 1 | 1 |
| 1 | 0 | 0 | 0 | 1 | 0 | 0 | 0 |
| 1 | 0 | 0 | 0 | 1 | 0 | 0 | 1 |
| 1 | 0 | 0 | 0 | 1 | 0 | 1 | 0 |
| 1 | 0 | 0 | 0 | 1 | 0 | 1 | 1 |
| 1 | 0 | 0 | 1 | 0 | 0 | 0 | 0 |
| 1 | 0 | 0 | 1 | 0 | 0 | 0 | 1 |
| 1 | 0 | 0 | 1 | 0 | 0 | 1 | 0 |
| 1 | 0 | 0 | 1 | 0 | 0 | 1 | 1 |
| 1 | 0 | 0 | 1 | 1 | 0 | 0 | 0 |
| 1 | 0 | 0 | 1 | 1 | 0 | 0 | 1 |
| 1 | 0 | 0 | 1 | 1 | 0 | 1 | 0 |
| 1 | 0 | 0 | 1 | 1 | 0 | 1 | 1 |

Upper group = port A and port C bits 4–7
Lower group = port B and port C bits 0–3

Figure 2.13   *The 8255 I/O board circuit*

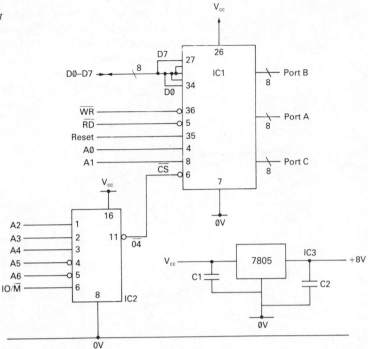

The circuit shown in Figure 2.13 used the chip as a general-purpose I/O device and will be used for most of the following I/O experiments and exercises.

D0 to D7 are connected to the chip as are the control ones $\overline{\text{WR}}$, $\overline{\text{RD}}$ and reset. A0 and A1 provide internal register selection as shown in Table 2.6. The address lines A2, 3, 4, 5and 6 are fed to the decoder 74 LS138 (IC2) and A7 is ignored. The 74 LS138 is also supplied with the IO/$\overline{\text{M}}$ signal. The decoder is active when A5 and A6 are low and IO/$\overline{\text{M}}$ is high. As output 4 is selected, then S3, S2 and S1 must be 1 0 0 respectively, so the address is given by:

| A7 | A6 | A5 | A4 | A3 | A2 | A1 | A0 |
|----|----|----|----|----|----|----|----|
| X  | 0  | 0  | 1  | 0  | 0  | X  | X  |

i.e. 10H–13H, and as A7 is ignored

90H–93H. By selecting other outputs of IC2 other I/O addresses could be assigned.

**Table 2.6**   *8255 register access*

| A1 | A0 | *8255 register* |
|----|----|-----------------|
| 0  | 0  | Port A data |
| 0  | 1  | Port B data |
| 1  | 0  | Port C data |
| 1  | 1  | Control |

To use this device it is necessary to program it for the purpose required. This is done by writing the control word to the control register at I/O address (13H). Refer to the data sheet for the required configuration.

### Exercise 2.7

Figure 2.14 is the sequence to program the port for all pins output.

Figure 2.14    *Programming the 8255 for output*

```
PASS 1

PASS 2

PASS 3
          0001 ;TO SET UP THE 8255 PPI FOR OUTPUT
          0002 ;ON ALL PORT (MODE 1) OPERATION.
          0003 ;
          0004 ;THE SYMBOLIC ADDRESSES FOR THIS
          0005 ;EXAMPLE ARE AS FOLLOWS, THEY MAY NEED
          0006 ;TO BE MODIFIED FOR YOUR ADDRESS
          0007 ;DECODING.
          0008 ;
3800      0009 ADAT:    EQU  50H      ;PORT A
3800      0010 BDAT:    EQU  51H      ;PORT B
3800      0011 CDAT:    EQU  52H      ;PORT C
3800      0012 PCON:    EQU  53H      ;CONTROL ADDRESS
          0013 ;
          0014 ;TO PROGRAM THE PORT WRITE THE
          0015 ;CONTROL WORD TO ''PCON''
          0016 ;
          0017 ;THE PROGRAM IS ASSEMBLE TO RUN
          0018 ;FROM ADDRESS 3800H
          0019 ;
3800      0020         ORG  3800H
          0021 ;
3800 3E80 0022 INIT:   MVI  A,80H    ;BIT 7=1 FOR CONTROL
3802 D353 0023         OUT  PCON     ;SEND TO PORT
          0024 ;
          0025 ;PORTS A, B & C ARE NOW READY TO
          0026 ;OUTPUT DATA USING THE ''OUT'' INSTRUCTION
          0027 ;E.G.
          0028 ;
3804 3EAA 0029         MVI  A,0AAH   ;AA = BINARY  1010 10
3806 D350 0030         OUT  ADAT     ;
          0031 ; ETC
3808      0032         END
ORIGIN; ENTRY = 3800; 3800
ADAT = 0050  BDAT = 0051  CDAT = 0052  PCON = 0053
INIT = 3800
```

Insert this sequence before the motor control program (Figure 2.12), then assemble and load this new program. Connect the motor interface to this I/O board and test this program.

Now let us examine some uses of the 8255 as an input device.

### Keyboard interface

Keyboard inputs to a processor system can be implemented in a number of ways. There are some very good dedicated hardware units

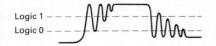

Figure 2.15 *A typical keybounce waveform*

available which debounce the keys and produce a binary or ASCII coded output, but for many small systems a software solution may be found.

The most difficult problem is that as the key is a mechanical device it does not make instant contact and stay put when the key is pressed, but rather produces a waveform as shown in Figure 2.15.

Therefore some method of 'debouncing' the key must be evolved. The most usual way is to examine the state of the key a few milliseconds after the first closure has been detected.

### Exercise 2.8

In this exercise three keys with pull-up resistors are assembled on a 'breadboard' and connected to PA0–PA2 of the 8255 (see Figure 2.16).

Figure 2.16 *Connecting three keys to the I/O board*

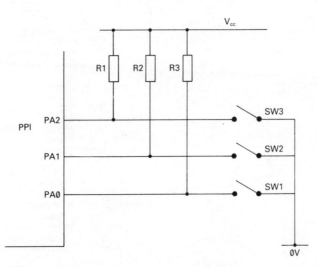

The program has to identify a key being pressed, debounce it and return with the key number in the A register.

Connection data is shown in Figure 2.16. A possible program is listed in Figure 2.17.

Enter, assemble and run this program.

Examine the A register to ensure that it contains the number of the key which was pressed.

How could the program be modified so that the processor could do other things as well as testing the state of the keyboard?

Figure 2.17  *Program to examine and debounce the keys*

```
              TITLE ''KYDET1''
       ;      PROGRAM :- KEY DETECT
       ;      VERSION :- 1.0
       ;
       ;
       ;      PROGRAMMER :- P.  D.  S.
       ;


       ;      DATE :- MAY 82
       ;
       ;
       ;      INPUT :- NOTHING
       ;
       ;
              OUTPUT :- NUMBER OF KEY PRESSED IN REG. A
       ;
       ;
              REGISTERS USED :- A, B, D, E, H, L.
       ;
       ;
       ;
       ;          PROGRAM DESCRIPTION
       ;          ===================
       ;
       ;      ROUTINE DETECTS AND DEBOUNCES SINGLE KEYS
       ;      CONNECTED TO BIT PA0-PA2 OF PORT A.
       ;      THE PROGRAM RETURNS WITH THE
       ;      NUMBER OF THE KEY PRESSED IN THE A REGISTER
       ;
       ;          TABLE OF EQUATES AND CONSTANTS
       ;          ==============================
PCON   EQU  53H
APOR   EQU  50H
DELY   EQU  057FH
STACK  EQU  3FFFH
MONW   EQU  57H
       ;
       ;          MAIN PROGRAM
       ;          ============
       ;
ENTER  LXI  SP,STACK      ;SET UP PROGRAM STACK
       MVI  A,90H         ;SET PORT FOR
       OUT  PCON          ; INPUT
       ;
       ;          GET INPUT AND TEST
       ;          ==================
PT     CALL GETKY         ;FIND INPUT
       JZ   INPT          ;NOT REALLY AN INPUT
       MOV  B,A           ;SAVE INPUT
       CALL DE10MS        ;DEBOUNCE KEY
       CALL GETKY         ;GET INPUT AGAIN
       CMP  B             ;STILL THE SAME?
       JZ   HITIT         ;YES GOOD KEY
       JP   INPT          ;NO, SO LOOK AGAIN
HITIT  CMP  4             ;IS INPUT=4?
       JZ   THREE         ;YES, SO KEY=3
       JP DONE            ;KEY=KEY NO.
THREE  MVI  A,3           ;LOAD A WITH KEY NUMBER
DONE   JP   MONW          ;RETURN-- THIS LINE CAN BE CHANGED
                          ; TO RET FOR SUBROUTINE USE
       ;          KEY COLLECTION ROUTINE
       ;          ======================
       ;
```

*continued*

```
GETKY      IN      APOR         ;GET INPUT
           ORI     0F8H         ;CLEAR UNUSED BITS
           CMA                  ;AND INVERT
           RET                  ;AND RETURN
;
;
;              DELAY ROUTINE
;              =============
;
;          NOTE:- NUMBERS IN BRACKET FOR EACH LINE
;                 ARE THE NUMBER OF CLOCK CYCLES FOR
;                 EACH INSTRUCTION.
;
DE10MS     PUSH    PSW          ;SAVE STATUS           (13)
           LXI     D,DELY       ;DELAY TIME            (10)
DE10       DCX     D            ;DECREMENT D           (06)
           MOV     A,D          ;CHECK IF FINISHED     (04)
           ORA     E            ;                      (04)
           JNZ     DE10         ;LOOP AGAIN            (07)
;                               ;TOTAL LOOP TIME=21 CYCLES
           POP     PSW          ;RESTORE STATUS
           RET                  ;AND RETURN
;
;          END OF DELAY ROUTINE
;
           END
```

## Exercise 2.9

Modify the program to guard against two or more keys being pressed at the same time. i.e. provide a measure of *roll-over*.

### A scanned matrix keyboard

If more than a few keys are required then it is usual to arrange them in a scanned matrix, as shown in Figure 2.18.

In this example, six lines are used to accommodate nine keys. Eight lines can handle sixteen keys and sixteen lines could handle sixty-four keys (8×8 matrix).

Figure 2.18   *A matrix keyboard*

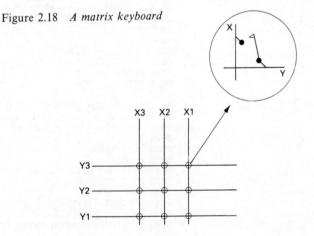

This example presents a software solution to the interfacing of a number of keys.

The operation of the software is to place a 0 on X1 and a 1 on all other lines, then examine the Y lines to see if a 0 appears on any of them. If not, then the 0 is moved to X2, X1 being returned to a 1 and the examination repeated. This continues until each line has been exercised when the sequence is repeated. If an 0 is found then the key is decoded with reference to a look-up table.

### Exercise 2.10

Figure 2.19 shows the connection arrangements. X0–X2 are connected to PA0–PA2 and Y0–Y2 are connected to PB0–PB2. Port A is set up for output and port B for input. Again, the keyboard could be set up on a breadboard or connected upon a stripboard such as Veroboard.

Figure 2.20 is the program listing for a program to scan the keyboard and return with the key which was pressed in the A register.

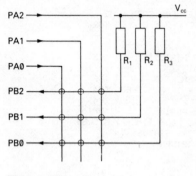

Figure 2.19  *Connecting a matrix keyboard*

Figure 2.20  *Program KEYSCAN*

```
0001 ;TITLE KEYSCAN
0002 ;VERSION 2.1
0003 ;JUNE 1982
0004 ;
0005 ;FUNCTION:- SUB-ROUTINE TO SCAN A MATRIX
0006 ;           KEYBOARD, DETECT AND DEBOUNCE
0007 ;           KEYSTROKES AND RETURN WITH A
0008 ;           CODE IN REGISTER A CONSISTING OF
0009 ;           THE UPPER NIBBLE = ROW AND THE
0010 ;           LOWER NIBBLE = COLUMN OF THE KEY
0011 ;           PRESSED.
0012 ;
0013 ;INPUT :- NOTHING
0014 ;OUTPUT :- CODE OF THE KEY PRESSED
0015 ;
0016 ;REGISTER USE :- A = RETURN OF KEY PRESSED
0017 ;                B = COLUMN COUNTER (3 IN THIS
0018 ;                EXAMPLE, MAY BE INCREASED TO
0019 ;                A MAXIMUM OF 64.)
0020 ;                C = CURRENT COLUMN POINTER.
0021 ;
0022 ;
0023 ;TABLE OF EQUATES ETC.
0024 APORT:EQU 50H
0025 BPORT:EQU 51H
0026 PCON:EQU 53H
0027 ;
0028 IO:EQU 82H ;PORT A - OUT, B - IN
0029 NKY:EQU 3 ;NUMBER OF COLUMNS
0030 IPAT:EQU 1 ;STARTING PATTERN FOR SCAN
0031 ;
0032   ORG 3100H
0033 ;
0034 INIT:MVI A,IO ;SET UP PORTS
0035   OUT PCON ,
0036 ;
0037 ;MAIN ROUTINE.
0038 ;
0039 KYDET:MVI A,NKY ;NUMBER OF COLS.
```

*continued*

```
0040   MOV B,A ;SET UP COL. COUNTER
0041   MVI A IPAT ;SET UP SCANNER
0042   MOV C,A ;
0043 MAIN:MOV A,C ;GET CURRENT COL.
0044   CALL TRY ;SEE IF KEY PRESSED
0045   JNZ GOTWUN ;YES
0046   MOV A,C ;SET UP NEXT COL.
0047   RLC ;
0048   MOV C,A ;AND SAVE
0049   DCR B ;ALL DONE?
0050   JNZ MAIN ;NO DO NEXT COL.
0051   JMP KYDET ;YES SO ROUND AGAIN OR
0052   ;              THIS MAY BE REPLACED BY RETURN
0053   ;
0054 TRY:CMA ;COMPLIMENT COL. COUNT
0055   OUT BPORT ;AND ENERGISE THE KEYBOARD
0056   IN APORT,GET ANY ROW
0057   CMA ;COMPLEMENT
0058   ORA A ;SET ZERO FLAG IF NO KEY
0059   RET ;AND RETURN
0060   ;
0061 GOTWUN:MOV D,A ;SAVE FIRST
0062   CALL WAI10 ;DELAY FOR DEBOUNCE
0063   CALL TRY
0064   CMP D ;SECOND=FIRST?
0065   JNZ  NOT ;NO
0066   RLC ;POSITION
0067   RLC ;TO UPPER
0068   RLC ;NIBBLE
0069   RLC ;
0070   ADD C ;ADD COLUMN POSITION
0071   RET ;RETURN TO CALLING PROGRAM
0072 NOT:JMP MAIN ;CARRY ON
0073   ;
0074 ;AN APPROXIMATE DELAY LOOP
0075 ;ACTUAL TIME CAN BE CHANGED BY
0076 ;LOADING A DIFFERENT NUMBER INTO
0077 ;THE HL REGISTER PAIR TO SUIT
0078 ;THE KEYS IN USE AS SOME HAVE
0079 ;LONGER BOUNCE TIMES THAN OTHERS
0080 WAI10:PUSH D ;SAVE REGISTERS
0081   PUSH H
0082   LXI  H,-1 ;
0083   LXI D,-1 ;
0084 WLUP:DAD D ;DEC HL
0085   JC WLUP ;
0086   POP H
0087   POP D
0088   RET
0089   END
```

### Exercise 2.11

Extend the program of Figure 2.20 to provide a decoding program such that the input is converted into a binary (hexadecimal) number in the A register corresponding to the key pressed.

One feature of a software-controlled keyboard is its flexibility. The key functions can be changed by simply changing the entries in the look-up table. Additionally, further keys can be added by extending the table and updating the key counter.

Now some (more) outputs.

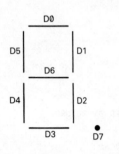

Figure 2.21 *Connecting a seven-segment display*

### Interfacing a seven-segment display

There are three main ways that seven-segment displays can be used by a microprocessor:

1   To treat each display as a memory or I/O locations and provide each display with its own data latch/decoder.
2   The display can be operated by a dedicated I/O driver chip.
3   The whole display can be under software control.

It is the last option which is presented here.

The assignment of data bits to segments is shown in Figure 2.21.

Table 2.7 shows the codes required to provide the display. In the circuit used, a logic 1 is used to turn a segment on and a 0 to turn it off. A similar arrangement is used for the display switching.

**Table 2.7**   *Seven-segment display codes*

| Display | D7 | D6 | D5 | D4 | D3 | D2 | D1 | D0 | Hex |
|---------|----|----|----|----|----|----|----|----|-----|
| 0 | 0 | 0 | 1 | 1 | 1 | 1 | 1 | 1 | 3F |
| 1 | 0 | 0 | 0 | 0 | 0 | 1 | 1 | 0 | 06 |
| 2 | 0 | 1 | 0 | 1 | 1 | 0 | 1 | 1 | 5B |
| 3 | 0 | 1 | 0 | 0 | 1 | 1 | 1 | 1 | 4F |
| 4 | 0 | 1 | 1 | 0 | 0 | 1 | 1 | 0 | 66 |
| 5 | 0 | 1 | 1 | 0 | 1 | 1 | 0 | 1 | 6D |
| 6 | 0 | 1 | 1 | 1 | 1 | 1 | 0 | 1 | 7D |
| 7 | 0 | 0 | 0 | 0 | 0 | 1 | 1 | 1 | 07 |
| 8 | 0 | 1 | 1 | 1 | 1 | 1 | 1 | 1 | 7F |
| 9 | 0 | 1 | 1 | 0 | 0 | 1 | 1 | 1 | 67 |
| A | 0 | 1 | 1 | 1 | 0 | 1 | 1 | 1 | 77 |
| B | 0 | 1 | 1 | 1 | 1 | 1 | 0 | 0 | 7C |
| C | 0 | 0 | 1 | 1 | 1 | 0 | 0 | 1 | 39 |
| D | 0 | 1 | 0 | 1 | 1 | 1 | 1 | 0 | 5E |
| E | 0 | 1 | 1 | 1 | 1 | 0 | 0 | 1 | 79 |
| F | 0 | 1 | 1 | 1 | 0 | 0 | 0 | 1 | 71 |
| . | 1 | 0 | 0 | 0 | 0 | 0 | 0 | 0 | 80 |

### Exercise 2.12

This demonstration program writes three numbers on to the display which are stored in three successive memory locations called DISBUF. The display consists of 3 seven-segment indicators.

Each number is displayed in turn for 10 ms. During set-up time it is necessary to blank the displays, otherwise the segments which are supposed to be off will be seen dimly flickering.

Figure 2.22 *Seven-segment display driver circuit*

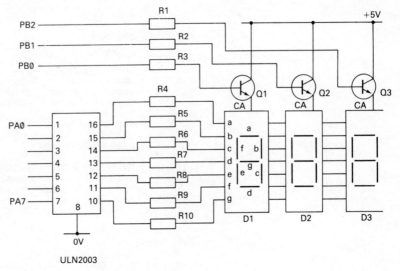

Figure 2.22 shows the hardware and connections necessary. Once again power interfacing is necessary. Q1–Q3 switch the current to the display, whilst IC1 provides the segment drive. The program is shown in Figure 2.23.

Figure 2.23 *Program multiplex DISPLAY*

```
;                          PROGRAM DESCRIPTION
;                          ===================

;          THIS MODULE DISPLAYS THE CONTENTS OF
;          A MEMORY BUFFER ON THREE SEVEN-SEGMENT
;          DISPLAYS CONNECTED TO THE I/O BOARD
;          SOFTWARE MULTIPLEXING OF THE DISPLAYS
;          IS USED.
;          THE NUMBERS STORED IN -DISBUF- SHOULD
;          BE WITHIN THE RANGE 00 TO 0F (HEX).

;          SOFTWARE CODE CONVERSION FROM HEX. TO
;          SEVEN SEGMENT CODE IS EMPLOYED.



;          INPUT: DIGITS FOR DISPLAY IN MEMORY ADDRSS 4000H

;          OUTPUT: NOTHING

;          USES REGISTERS:- A, B, C, D, H & L

;                    TABLE OF EQUATES AND CONSTANTS
;                    ===============================
;
APORT     EQU        50H      ;PORT ADDRESSES
BPORT     EQU        51H
CPORT     EQU        52H
PCON      EQU        53H
;
DELY      EQU        057FH    ;DELAY CONSTANT FOR DISPLAY
```

*continued*

```
;
                PAGE
;
                ORG   3000H
;
START   MVI     A,80H     ;SET UP ALL PORTS FOR OUTPUT
        OUT     CPORT     ;
        XRA     A         ;CLEAR ACCUMULATOR
        OUT     APORT     ;TURN OFF
        OUT     BPORT     ;DISPLAYS
;
;               END OF INITIALIZATION
;
;               MAIN PROGRAM LOOP
;
LOOP    MVI     B,3       ;NUMBER OF DISPLAYS
        MVI     C,4       ;DISPLAY POINTER
        LXI     H,DSBUF   ;POINT TO START OF DISPLAY BUFFER
LOOP2   MOV     A,M       ;GET MOST SIG. DIGIT



        CALL    CONVT     ;CONVERT TO SEVEN SEG. CODE
        OUT     APORT     ;AND PREPARE DISPLAY
        MOV     C,A       ;SWITCH ON
        OUT     BPORT     ;THE DISPLAY
        CALL    DE10MS    ;WAIT FOR 10MS
        XRA     A         ;CLEAR ACC.
        OUT     BPORT     ;AND TURN OFF DISPLAY
        INX     H         ;POINT TO NEXT NUMBER
        MOV     A,C       ;MOVE DISPLAY POINTER ON
        RRC               ;TO NEXT
        MOV     C,A       ;
        DCR     B         ;DECREMENT LOOP COUNT
        JNZ     LOOP2
        JP      LOOP      ;CONTINE FOR EVER

                END OF MAIN LOOP
CONVT   PUSH    H         ;SAVE CALLING
        PUSH    B         ;PROGRAM STATUS
        MVI     C,0       ;CLEAR C
        LXI     H,TABLE   ;TOP OF LOOKUP TABLE
        ADD     L         ;A=TABLE+NUMBER
        MOV     L,A       ;TOTAL TO L-REG
        MOV     A,H       ;GET UPPER BYTE
        ADC     C         ;AND CHECK FOR CARRY
        MOV     H,A       ;RESTORE H-REG
;
;               HL IS NOW POINTING AT THE EQUIVALENT SEVEN
;               SEGMENT CODE
;
        MOV     A,M       ;GET CODE
        POP     B         ;RESTORE
        POP     H         ;STATUS
        RET               ;AND RETURN
;
;               END OF CONVERSION ROUTINE
;
                PAGE
;
;                   DELAY ROUTINE
;                   =============
;
;               NOTE:- NUMBERS IN BRACKET FOR EACH LINE
;                      ARE THE NUMBER OF CLOCK CYCLES FOR
;                      EACH INSTRUCTION.
;
```

*continued*

```
DE10MS    PUSH    PSW       ;SAVE STATUS        (13)
          LXI     D,DELY    ;DELAY TIME         (10)
DE10      DCX     D         ;DECREMENT D        (06)
          MOV     A,D       ;CHECK IF FINISHED  (04)
          ORA     E         ;                   (04)
          JNZ     DE10      ;LOOP AGAIN         (07)
;                           ;TOTAL LOOP TIME=21 CYCLES
          POP     PSW       ;RESTORE STATUS
          RET               ;AND RETURN
;
;         END OF DELAY ROUTINE
;
;
;
;
;                           LOOKUP TABLE OF CODES
;                           =====================
;
TABLE     BYTE    3FH,06,5BH,4FH,66H,6DH
          BYTE    7DH,07,7FH,67H,77H
          BYTE    7CH,39H,5EH,79H,71H
;
;
;
;                           PROVIDE DISPLAY BUFFER
;                           ======================
;
          ORG     4000H
;
DSBUF     BLOCK   3         ;RESERVE THREE BYTES
;
;
          END               ;END OF PROGRAM
```

### Exercise 2.13

As the program in Figure 2.23 stands, there is no protection against invalid codes if the number in DISBUF is in excess of 0FH.

Add protection to the program such that if the number is out of range 00–0FH is displayed.

### Exercise 2.14

Combine the programs of Figures 2.20 and 2.23 to ECHO the numbers typed on the keyboard on to the display and to shift the numbers left as new ones are typed in so that the new number always appears on the rightmost digit.

### Analogue interfacing

Many quantities which need to be measured and/or reduced are analogue quantities. As the microprocessor only deals with numbers, these analogue quantities have to be converted to a numerical representation before processing by the computer. This function is performed by an analogue to digital converter.

*Analogue to digital converter (ADC)*
Two main types are used with microprocessors:

1  The counter type.
2  The successive approximation type.

The latter is more convenient in microprocessor systems as it is at least an order of magnitude faster than the former.

The heart of an analogue to digital converter is usually a *digital to analogue converter* (DAC).

### Exercise 2.15   The DAC

A digital to analogue converter provides an output voltage proportional to the supplied binary input. Figure 2.24 shows the circuit of a simple 4-bit DAC.

Figure 2.24   *A four-bit DAC*

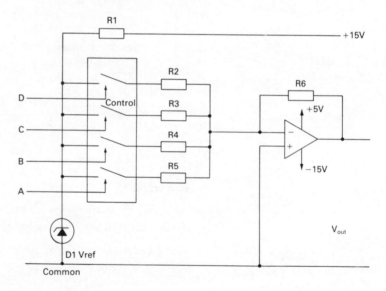

Figure 2.25   *DAC driver program*

```
0001 ;TITLE :- 4-BIT DAC
0002 ;VERSION 1.0
0003 ;
0004 ;DATE APRIL 1982
0005 ;
0006 ;INPUT :- NOTHING
0007 ;
0008 ;OUTPUT :- NOTHING
0009 ;
0010 ;REGISTERS USED :- ALL
0011 ;
0012 ;         PROGRAM DESCRIPTION
0013 ;         ===================
0014 ;
0015 ;THIS IS A DEMONSTRATION PROGRAM TO
0016 ;DRIVE A SIMPLE 4-BIT D - A CONVERTER.
0017 ;AN INPUT IS TAKEN FROM THE KEYBOARD,
```

*continued*

```
0018 ;CONVERTED INTO A 4-BIT BINARY NUMBER,
0019 ;AND OUTPUT TO THE D - A WHICH IS CONNECTED
0020 ;TO BITS PA0 - PA3 OF THE 8255 PERIPHERIAL
0021 ;EXTENSION BOARD.
0022 ;ANY NON-HEX CHARACTER WILL TERMINATE THE PROGRAM.
0023 ;
0024 ;
0025 ;TABLE OF EQUATES AND CONSTANTS
0026 ;
0027 ;MONITOR CALLS
0028 ;
0029 MONW:EQU 57H
0030 NEWL:EQU 2DAH
0031 KB:EQU 5BEH
0032 KLUC:EQU 660H
0033 VDU:EQU 6C0H
0034 ;
0035 ;I/O ADDRESSES
0036 ;
0037 APOR:EQU 50H
0038 PCON:EQU 53H
0039 ;
0040 ORG 3800H
0041 ;
0042 ;START OF PROGRAM
0043 ENTER:MVI A,80H ;SET UP PORT
0044  OUT PCON ;
0045 ;
0046 RUN:CALL NEWL ;CRLF
0047  CALL KB ;GET NUMBER
0048  CALL HEXCON ;CONVERT TO BINARY
0049  OUT APOR ;OUTPUT TO DAC
0050  JMP RUN ;AND WAIT FOR THE NEXT
0051 ;
0052 ; END OF MAIN PROGRAM
0053 ;
0054 ;
0055 ;    ASCII TO BINARY CONVERSION
0056 ;ENTER THIS ROUTINE WITH THE ASCII CHARACTER
0057 ;FOR CONVERSION IN THE A - REGISTER
0058 ;ROUTINE RETURNS WITH CONVERTED NUMBER
0059 ;IN THE A - REGISTER OR, IF OUT OF RANGE
0060 ;THEN JUMPS TO MONITOR.
0061 ;
0062 HEXCON: CALL KLUC ; CONVERT TO LOWER CASE
0063  SUI 30H ;A=A-48
0064  JM ERR ;NOT HEX
0065  CPI 6AH ;A 10?        (ED: NO CHEVRON)
0066  JC GOTIT
0067  CPI 11H ; A 16?       (ED : NO CHEVRON)
0068  JC ERR ;
0069   CPI 17H ;NUMBER F?   (ED: NO CHEVRON)
0070  JNC ERR
0071  SUI 7 ;CONVERT RANGE A-F
0072 GOTIT: RET ; RETURN O.K.
0073 ERR:JMP MONW ;NOT HEX NUMBER
0074 END
```

Connect up the circuit, as shown, and use PA0 to PA3 to drive the inputs. Connect a voltmeter across the output and use the program of Figure 2.25 to set up different values on the DAC input lines. Record the outputs. Plot a graph of output against binary input.

*Note*: Only the keyboard input 0 to 9 and A to F are valid – any other value will stop the program and return to the monitor.

*Questions*

1  What is the full-scale output of this simple circuit?
2  What is the resolution and what is its relation to $V_{ref}$ ?

*An 8-bit DAC*

The ZN425E is an 8-bit D/A to A/D integrated circuit. Figure 2.26 shows this connected to port A of the PPI. By setting the maximum output (all inputs high) to 2.56 V, a resolution of 0.01 V or 10 mV per bit can be obtained. Similarly, any other scale factors could be applied to give different resolutions by adjusting the amplification of the output buffer amplifier.

Figure 2.26  *The ZN425E*

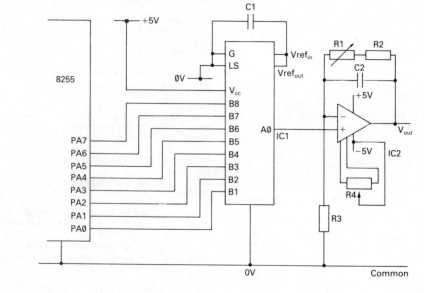

*Exercise 2.16  Waveform generation*

A typical application for a DAC is to generate a specific output waveform to interface with a power controller.

Figure 2.27 shows a program to generate a triangle wave output. The period of the waveform is determined by the time delay given to each output step. Changing the constant PERD: will allow different periodic times.

More complex patterns can be generated but will require a look-up table or calculations by the processor.

Connect a CRO to the output of the ZN425E. Enter, assemble and run the program. Try different values of PERD: and note the effect on the frequency of the wave.

Figure 2.27   *Program TRIANGLE*

```
0001 ;TITLE TRIANGLE
0002 ;
0003 ;GENERATES A TRIANGLE (SAWTOOTH)
0004 ;WAVEFORM VIA A D-A CONVERTER CONNECTED
0005 ;TO PORT A OF THE 8255 EXPANSION BOARD.
0006 ;
0007 ;INPUT - NOTHING
0008 ;OUTPUT - NOTHING
0009 ;
0010 ;
0011 PORT:EQU 0CH
0012 PCON:EQU 0FH
0013 ;
0014 TIME:EQU 80H
0015 ;
0016   ORG 3800H
0017 ;
0018 INIT:MVI A,80H ;SET UP PORT FOR
0019   OUT PCON ;OUTPUT
0020   XRA A ;CLEAR A
0021 ;
0022 ;GENERATE UP SLOPE OF RAMP
0023 ;
0024 UP:OUT PORT ;SET VALUE TO DAC
0025   CALL PAUSE ;INCREMENTAL TIME
0026   INR A
0027   JNZ UP ;IF GREATER THAN ZERO THEN CONTINUE
0028 DOWN:OUT PORT ;NOW RAMP DOWN
0029 CALL PAUSE ;
0030   DCR A ;GOING DOWN
0031   JNZ DOWN ;LESS THAN ZERO CONTINUE
0032   JMP UP ;GO BACK UP
0033 ;
0034 PAUSE:PUSH PSW;SAVE STATUS
0035   MVI B,TIME ;SET DELAY
0036 PAUSE2:DCR B ;DECREMENT
0037   JNZ PAUSE 2;
0038   POP PSW ;RESTORE FLAGS
0039   RET ;DONE
0040   END ;
```

### Exercise 2.17   *Sinewave generation*

Write a program to output successive values to the DAC such that a
sinewave is generated. Figure 2.28 provides a reasonably coarse look-
up table and it assumes that the output waveform will have a
peak–peak amplitude of 2 V.

Figure 2.28   *Program SIN*

```
0001 ;TITLE SIN
0002 ;
0003 ;GENERATES A SINEWAVE FROM A LOOK-UP
0004 ;TABLE VIA A D-A CONVERTER CONNECTED
0005 ;TO PORT A OF THE 8255 EXPANSION BOARD.
0006 ;
0007 ;INPUT - NOTHING
0008 ;OUTPUT - NOTHING
0009 ;
0010 ;
0011 PORT:EQU 0CH
0012 PCON:EQU 0FH
0013 ;
0014 TIME:EQU 80H
```

*continued*

```
0015 ;
0016  ORG 3800H
0017 ;
0018 INIT:MVI A,80H ;SET UP PORT FOR
0019  OUT PCON ;OUTPUT
0020 ;
0021 UP:LXI H,TAB ;START OF LOOKUP
0022  MVI C,LEN ;SET ENTRY COUNTER
0023 UP1:MOV A,M ;GET FIRST VALUE
0024  OUT PORT ;AND SET IT
0025  CALL PAUSE ;INCREMENTAL TIME
0026  INX H ;POINT TO NEXT ENTRY
0027  DCR C ;DOWN COUNTER
0028  JNZ UP1;CONTINUE IF NOT UP
0029 DOWN:MVI C, LEN ;RELOAD COUNTER
0030 DOWN2:DCX H ;RETURN DOWN THE TABLE
0031  MOV A,M ;GET THE VALUE
0032  OUT PORT ,
0033  CALL PAUSE ;
0034  DCR C ;DOWN COUNTER
0035  JNZ DOWN2;LESS THAN ZERO CONTINUE
0036 JMP UP ;GO BACK UP
0037 ;
0038 PAUSE:PUSH PSW;SAVE STATUS
0039  MVI B,TIME ;SET DELAY
0040 PAUSE2:DCR B ;DECREMENT
0041  JNZ PAUSE2 ;
0042  POP PSW ;RESTORE FLAGS
0043  RET ;DONE
0044 ;
0045 ;TABLE OF VALUES
0046 ;
0047 TAB:DB  0
0048  DB 5
0049  DB 10
0050  DB 15H
0051  DB 25H
0052  DB 39H
0053  DB 4FH
0054  DB 65H
0055  DB 80H
0056  DB 98H
0057  DB 00B0H
0058  DB 0C6H
0059  DB 0DAH
0060  DB 0EAH
0061  DB 0F5H
0062  DB 0FCH
0063  DB 0FFH
0064 LEN:EQU LEN-TAB ;LET ASSEMBLER SORT OUT
0065 ;                    LENGTH OF TABLE.
0066  END
```

### Exercise 2.18

Modify the program and look-up table to provide a less coarse waveform.

### Analogue to digital conversion

With the addition of simple logic, the DAC can be modified to perform the A–D function.

Figure 2.29 shows the block diagram of a typical counter-type A–D converter system, using the ZN425E.

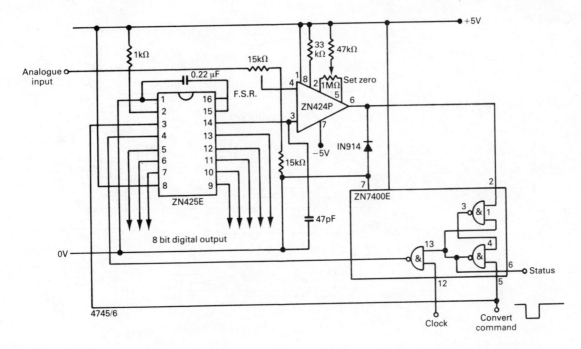

Figure 2.29 *Analogue to digital converter*

When the 'convert command' is issued, a number of things happen:

1  The counter is reset to zero.
2  The status line is put to FALSE.
3  The clock is gated to the counter which then starts to increment.

The output of the counter face is fed to the DAC, the output of which is used as one input to the comparator. When the DAC output is equal to the analogue input, the comparator output changes state and the following sequence occurs:

1  Disconnect clock from counter.
2  Put status line TRUE.

A digital value can now be read on the output lines of the converter.

### Exercise 2.19  *Interfacing a counter-type ADC*

Figure 2.30 shows the addition of a ZN425E ADC to the I/O board. The CA3130 is used as a comparator and the 7400 as control logic. Refer to the data sheet in Appendix 2 for full operational details.

The data is input to the system via port A and port C is used in its 'handshake' mode to control the ADC. A 'start conversion' pulse is sent from pin PC5 to the ADC logic by reading the data register.

Figure 2.30   *The ZN425E as an ADC*

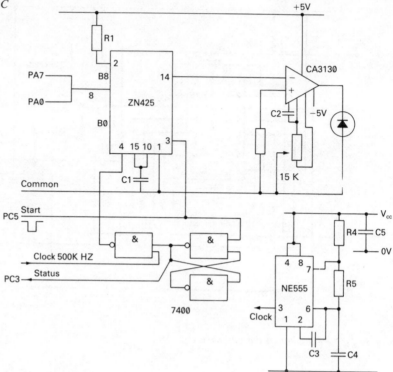

The program then polls the input for the 'conversion complete' signal from pin PC4 of port C.

This means operating the PPI in mode 1 (strobed input/output). This splits port C into two 'handshake' control sections. PC0–3 are associated with port A and PC4–7 with port B.

The assignments of these pins are shown in Table 2.8.

**Table 2.8**   *Port C usage in mode 1*

| Bit | D7 | D6 | D5 | D4 | D3 | D2 | D1 | D0 |
|---|---|---|---|---|---|---|---|---|
| Input | I/O | I/O | IBF A | $\overline{\text{STRA}}$ | INT A | $\overline{\text{STR}}$ B | IBF B | INT B |
| Output | $\overline{\text{OBF}}$ A | $\overline{\text{ACK}}$ A | I/O | I/O | INT A | $\overline{\text{ACK}}$ B | $\overline{\text{OBF}}$ B | INT B |

*Port C signal description*

INT   Interrupt request (port A or B).

This pin is set to logic 1 by the operation of the $\overline{\text{STR}}$ pin. In the input mode it is reset by an IN PORT operation and in output mode it is reset by an OUT PORT operation. This pin can be used to interrupt the CPU.

$\overline{\text{STR}}$    Active low input latches data into input buffer on the port – sets INT high and sets IBF high.

IBF    Indicates that data is latched into the input buffer and is available for transfer to the bus. This pin is reset by an IN PORT operation.

$\overline{\text{OBF}}$    Output equivalent of IBF. This pin goes low as a result of an OUT PORT operation to write data to a peripheral.

$\overline{\text{ACK}}$    Active low input. The receiving devices pull this pin low to inform the PPI that it has read the data sent.

The state of any of these pins can be examined by the CPU by an IN PORT C status instruction. The bit arrangement is shown in Table 2.9.

**Table 2.9**    *Status word bit arrangement*

| *Bit* | D7 | D6 | D5 | D4 | D3 | D2 | D1 | D0 |
|---|---|---|---|---|---|---|---|---|
| *Input* | I/O | I/O | IBF A | INT A | INTE A | INTE B | IBF B | INT B |
| *Output* | $\overline{\text{OBF}}$ A | INTE A | I/O | I/O | INT A | INTE B | $\overline{\text{OBF}}$ B | INT B |

The bits labelled INTE are the interrupt enable flags for the ports. When INTEx is a logic 1 the INT pin is enabled and when INTEx is logic 0 the pin is disabled.

These bits are manipulated with the bit set/reset facility. This is illustrated in Figure 2.31.

Figure 2.31    *Bit set/reset on the 8255*

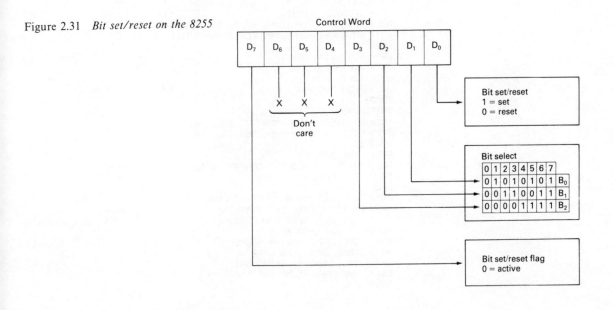

To operate the ADC, the processor system must provide a 'start conversion' signal, wait for a 'conversion complete' signal, then read the data from port A. So port A is programmed for input, port B for output and port C as the control lines. (Port B control is used for 'start conversion'.)

Connect the ADC data lines to port A, the status to $\overline{\text{STB A}}$ and the convert command to the $\overline{\text{OBF B}}$ pins. Connect $\overline{\text{DBF B}}$ to $\overline{\text{ACK B}}$; this will produce a plus on $\overline{\text{OBF B}}$ of one clock cycle for ADC initiation.

The command word for the PPI in this mode is:

1   0   1   1        1   1   0   X
=(BC HEX)

### Exercise 2.20   Calibration of the ADC

Connect the ADC to the peripheral board as shown in Figure 2.32. Connect the input to a variable voltage source which can be set accurately between 0 V and 2.56 V.

The following program converts the input and outputs the value on the screen in hex, with a short pause between each sample.

Assemble, load and run the program of Figure 2.32.

Figure 2.32   *Program ADCCON*

```
0001 ;TITLE ADC CON
0002 ;
0003 ;ANALOG TO DIGITAL CONVERTER
0004 ;DRIVER PROGRAM
0005 ;
0006 ;THIS PROGRAM WILL TRIGGER THE ADC TO
0007 ;START THE CONVERSION, THEN WAIT FOR
0008 ;THE CONVERSION TO OCCUR, THEN
0009 ;DISPLAY THE RESULTANT VALUE AS
0010 ;A HEX. NUMBER
0011 ;THE PROGRAM LOOPS CNTINUOUSLY WITH
0012 ;A SHORT PAUSE BETWEEN CONVERSIONS.
0013 ;
0014 PORTA:EQU 80H
0015 PORTB:EQU 81H
0016 PORTC:EQU 82H
0017 PCON:EQU 83H
0018;
0019 MONDEL:EQU 732H
0020 DISA:EQU 35DH ;MONITOR CALL PRB
0021 CRLF:EQU 2DAH ;MONITOR CALL PRNL
0022 ;
0023  ORG 3800H
0024 INIT: MVI A,0BCH ;CONTROL WORD
0025  OUT PCON ;
0026 ;
0027 CONV:OUT PORTB ;START CONVERSION
0028 WOT: IN PORTC ;GET STATUS
0029  ANI 10H ;TEST FOR READY
0030  JZ WOT ;NOT READY
0031  CALL DISA ;PRINT VALUE
0032  CALL CRLF ;NEW LINE
0033  CALL MONDEL ;SHORT WAIT
0034  JMP CONV ;CONTINUE FOR EVER
0035  END
```

Set the input to 2.56 V and adjust the 'set FSR' pot for an output of FFH ±1 bit.

Set the input to 0 V and check the reading is 00 ±1.

Set the input to about 20 different values between these limits and obtain a graph of $V_{in}$ against reading. The correspondence should be 1 mV/bit.

*Question*

Is the $V_{in}$/number out relationship linear for this converter?

### Exercise 2.21

The converter now has a full-scale range (FSR) of 2.5 V with an input resistance of 15 kΩ. If it is required to measure higher voltages then multiplier resistors must be added in series with the input. Similarly, if lower voltages are required then some amplification will be needed before the input resistor.

Figure 2.33 shows a set-up for reading 0 to 25.6 V with a resolution of 100 mV. Use this arrangement to measure some voltages in a piece of electronic equipment.

Figure 2.33  *Multiplying the input of an ADC*

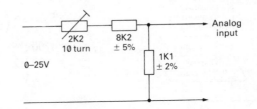

In many cases the input voltage for an ADC comes from some type of transducer such as a strain gauge bridge, a pressure transducer or a thermistor.

### Exercise 2.22  *The digital thermometer*

This exercise uses the ADC to input voltage from a thermistor network, correct non-linearity and display the temperature measured on the screen of the VDU.

The ADC conversion routine is the same as that shown in Figure 2.32.

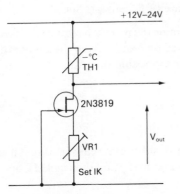

Figure 2.34  *Thermistor constant current circuit*

Connect the circuit to the specification of Figure 2.34. Allow the component temperature to stabilize; this may take a minute or so. Run the program of Figure 2.32 and read the output voltage. Measure the ambient temperature using a normal thermometer.

Using the data sheet for your thermistor (see Appendix 2 for suggested thermistor), find the resistance corresponding to the ambient temperature and calculate the correct voltage which should be across the thermistor for a current of 0.15 mA.

*Example*

$$R_{298} = 4700 \ \Omega$$
$$B = 3390$$
$$I_{th} = 0.15 \ mA$$

If $t(amb) = 290$ K (17 °C), then

$$R_{(290)} = R_{(298)} \exp \left( \frac{B}{290} - \frac{B}{298} \right)$$

$$= 4700 \exp \left( \frac{3390}{290} - \frac{3390}{298} \right)$$

$$= 4700 \ e^{0.31}$$

$$= 6408 \ \Omega$$

This corresponds to a voltage of

$$6408 \times 0.15 \times 10^{-3}$$
$$= 0.96 \ V \qquad\qquad (1)$$

at 0° C,

$$R_{273} = R_{298} \exp \left( \frac{3390}{273} - \frac{3390}{298} \right)$$

$$= 4700 \ e^{1.04}$$

$$= 13297 \Omega$$

$$= 1.99 \ V \qquad\qquad (2)$$

If value (1) is not obtained, then adjust R1 for this voltage.

Now immerse the thermistor in iced water and check that the equivalent voltage for a temperature of 0 °C is produced.

Having set up the circuit, let us examine the program requirements to allow the temperature to be displayed on request from the keyboard. Figure 2.35 shows a flow chart for a possible program.

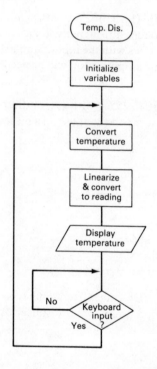

Figure 2.35 *Flow chart SHOWTEMP*

### Exercise 2.23

Write a program to satisfy this flow chart.

This program lends itself to a series of subroutine calls in as much as each can be assembled and tested individually and then called from a short main program.

The display of temperature should use routines which are already in the HEKTOR monitor. PRMES at location 030AH has already been used in previous programs – this should be arranged to write the text 'temperature is' to the screen. The digits can be printed by the routine PRBI at location 0359H. This routine takes the hex numbers in locations specified by the HL register and prints them on the screen. If a decimal readout is required, then the program has to convert the hex to decimal before storing them in T BUFF, the buffer for PRBI.

### More than one input

Sometimes it is required to input more than one analogue variable into the microprocessor system. In these cases each input (or channel) could have its own ADC, but this could prove expensive, especially when high-accuracy ADCs are used. A method of resolving this problem is to multiplex the analogue channels into the ADC. Suitable devices are CD4051B, CD4052B or CD4053B.

The CD4051 is a single eight-channel analogue multiplexer. The internal circuit is shown in Figure 2.36. The multiplexing action is performed by CMOS switches, TG1 to TG8. When the appropriate control line is activated the input pin C0–C7 is connected to the common out pin (3). Which control line is active is determined by the binary code on the 'select' input pins (9, 10 and 11). A further inhibit line is provided (pin 6) but is not used in this application.

Figure 2.36 *Block diagram 4051 multiplexer (source:* RCA Data Book (CMOS)

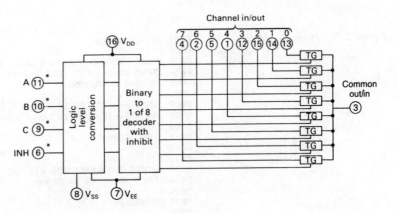

Note that as TG1 to TG8 behave as switches, it is possible to use the device to switch a voltage from the common 'output' to each input, if required.

Each switch element is not a 'perfect' switch. When in its 'on' condition, it represents a resistance of between 470 Ω and 1050 Ω at 25 °C for a 5 V supply, and this must be taken into account if any multiplier resistors are used before the ADC.

Similarly, in the 'off' condition a leakage current of up to 100 nA may be encountered.

### Exercise 2.24   *Multiplexing the input to an ADC*

The connections for this exercise are shown in Figure 2.37.

To select which channel is to be read, Port B is connected to the 'select' input on the multiplexer. The program selects the channel to be read and then a conversion sequence is initiated. The program is shown in Figure 2.38.

Figure 2.37   *Connecting a multiplexer to the ADC*

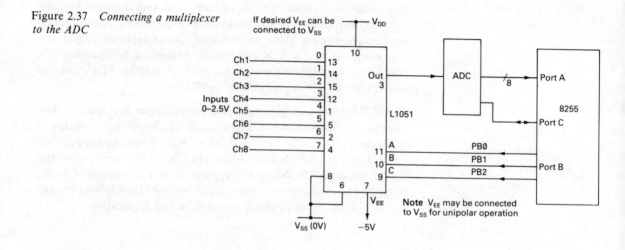

Figure 2.38   *Program 8IN*

```
0001 ;TITLE INSCH
0002 ;
0003 ;THIS PROGRAM INPUTS 8 VOLTAGE READINGS
0004 ;FROM AN 8-INPUT MULTIPLEXER INTO SUCCESSIVE
0005 ;MEMORY LOCATIONS.
0006 ;
0007 ;THIS PROGRAM CALLS : - ADC CON.
0008 ;AN MODIFIED ADC DRIVER PROGRAM.
0009 ;
0010  ORG 3800H
0011 ;
0012 ;
0013 ;OTHER I/O ADDRESSES ETC ARE TO BE ADDED HERE
0014 ;FROM PREVIOUS PROGRAMS
0015 ;
0016 CHANOW:DS 1
0017 ;
0018  ORG 3801H
0019 ;
0020 START:MVI A 0BCH
0021  OUT PCON;SET UP PORT
0022  MVI A,-1 ;CHAN 0-1
0023  STA CHANOW ;INIT CHANNEL POINTER
0024  LXI H,BUFFA ;SET UP RECEIVING AREA
0025 ;
0026 ;MAIN RUN LOOP
0027 ;
0028 GETTEM:MVI B,8 ;# OF CHANS.
0029 GETEM2: LDA CHANOW ;GET CHANNEL # - 1
0030  INR A ;SET TO CURRET
0031  STA CHANOW ;SAVE IT
0032  PUSH B ;USED IN CONVERSION ROUTINE
```

*continued*

```
0033  CALL ADCON ;CONVERT. RETURNS IN A
0034  MOV A,M ;SAVE VALUE
0035  INX H ;NEXT BUFFER
0036  DCR B
0037  JNZ GETEM2
0038  JMP MONW
003⊃ ;
0040  BUFFA:DS 8 ;STAORGE FOR DATA
0041  END
```

Note that this program calls routines previously developed and is not complete – if your development system supports a program library (Ch1) then the routines could well be stored there.

This program selects each input in turn to be applied to the input of the ADC. A convert command is then issued and a delay loop entered which polls the ADC until it has completed the conversion. The input is then stored in a table in memory.

This type of program could be used, for example, to read test voltages in an automatic test situation.

Enter, assemble and load the program. Then provide a number of different voltages in the range 0 to 2.5 V to at least three inputs. Run the program and then examine the memory locations to check that the correct values have been stored.

### Exercise 2.25

Modify the above program to provide a readout on the monitor screen of the difference between the measured value of voltage and the correct value, which has been previously stored in memory. A further refinement would be to print a warning message if the measured value was outside a set tolerance.

### Driving stepper motors

The stepper motor has gained popularity owing to the fact that it can be used to drive mechanical devices with precision and without the use of position-sensing transducers, i.e. effective open loop control.

Stepper motors fall into three main types:

1   Permanent magnet.
2   Variable reluctance.
3   Hybrid.

The stepper motor requires its coils to be energized in a specific sequence if the rotor is to rotate and provide useful torque.

Consider the two-coil motors shown in Figure 2.39. If the switches are operated in the sequence shown then the motor will rotate; reversing the sequence causes the direction of rotation to reverse.

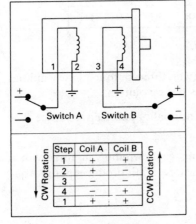

Figure 2.39 *Stepper motor connections*

| Step | Coil A | Coil B |
|------|--------|--------|
| 1 | + | + |
| 2 | + | − |
| 3 | − | − |
| 4 | − | + |
| 1 | + | + |

This bipolar drive system can be implemented with the push–pull circuit shown in Figure 2.40. In some cases it is easier to provide unipolar drive as shown in Figure 2.41. This provides less torque at low speed but the performance is about equal at higher speed.

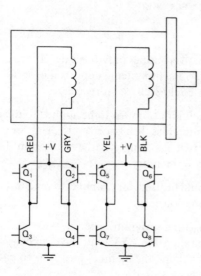

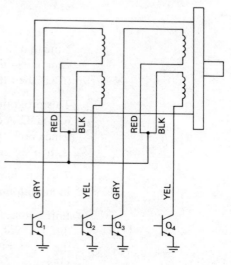

Bipolar

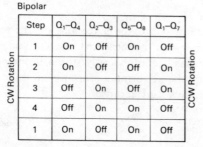

| | Step | $Q_1$–$Q_4$ | $Q_2$–$Q_3$ | $Q_5$–$Q_8$ | $Q_1$–$Q_7$ | |
|---|---|---|---|---|---|---|
| CW Rotation | 1 | On | Off | On | Off | CCW Rotation |
| | 2 | On | Off | Off | On | |
| | 3 | Off | On | Off | On | |
| | 4 | Off | On | On | Off | |
| | 1 | On | Off | On | Off | |

Figure 2.40   *Push–pull drive*

| | Step | $Q_1$ | $Q_2$ | $Q_3$ | $Q_4$ | |
|---|---|---|---|---|---|---|
| CW Rotation | 1 | ON | OFF | ON | OFF | CCW Rotation |
| | 2 | ON | OFF | OFF | ON | |
| | 3 | OFF | ON | OFF | ON | |
| | 4 | OFF | ON | ON | OFF | |
| | 1 | ON | OFF | ON | OFF | |

Figure 2.41   *Unipolar drive*

Stepper motors are available with two-, three- or four-phase windings and with 2 steps to 200 plus steps per revolution.

Step sequence can be wave drive, normal or half-step sequence. The latter two are explained in Table 2.10.

Wave drive consists of energizing each winding in turn with a 'walking one' output waveform. However, this produces poor torque and is not popular.

*Sequence driving*

A motor of the form shown in Figure 2.41 can be driven by the unipolar four- or eight-step sequence, shown in Table 2.10.

**Table 2.10** *Sequence driving of stepper motor*

(a) *Four-step sequence*

|  | Winding | | | |
|---|---|---|---|---|
| Step | 1 | 2 | 3 | 4 |
| 1 | on | off | on | off |
| 2 | on | off | off | on |
| 3 | off | on | off | on |
| 4 | off | on | on | off |
| 1 | on | off | on | off |

clockwise rotation (left) — counter-clockwise rotation (right)

(b) *Eight-step sequence*

|  | Winding | | | |
|---|---|---|---|---|
| Step | 1 | 2 | 3 | 4 |
| 1 | on | off | on | off |
| 2 | on | off | off | off |
| 3 | on | off | off | on |
| 4 | off | off | off | on |
| 5 | off | on | off | on |
| 6 | off | on | off | off |
| 7 | off | on | on | off |
| 8 | off | off | on | off |
| 9 | on | off | on | off |

clockwise rotation (left) — counter-clockwise rotation (right)

These sequences produce about 30 per cent more torque than wave drive; the eight-step sequence doubles the number of steps per revolution, so halving the rotation angle for each step and therefore increasing the precision.

The program for this drive is, by definition, more complex than for simple wave driving.

### Exercise 2.25 *Driving a 200 step revolution motor*

The program presented in Figure 2.42 provides a suitable sequence to drive the stepper motor a number of degrees depending upon the number loaded into the B register. For this, one revolution is divided into 200 steps, or 360/200 degrees/step = 1.8°/step.

Using half-step drive, this reduces to 0.9°/step.

Hence the position of the shaft can be controlled by feeding the motor with the appropriate number of steps.

The sequence of steps is contained in the table STAB: and is accessed by the HL register.

Figure 2.42   *Program ONESTEP*

```
0001 ;TITLE- STEP1
0002 ;
0003 ;VERSION 1.2
0004 ;
0005 ;FUNCTION:- THIS ROUTINE DRIVES A 200 STEP/REV.
0006 ;           STEPPER MOTOR A GIVEN NUMBER OF STEPS.
0007 ;           THE RESULTANT ROTATION WILL BE
0008 ;           1.8*N DEGREES WHERE N IS THE NUMBER
0009 ;           LOADED INTO THE B - REGISTER AT THE
0010 ;           START OF THE ROUTINE
0011 ;
0012 ;INPUT - INT(DEGREES/1.8) IN THE B-REGISTER
0013 ;
0014 ;OUTPUT - NOTHING
0015 ;
0016 ;USES REGISTERS A-B-C & HL
0017 ;
0018 ;EQUATES ETC.
0019 ;
0020 APORT:EQU 0CH
0021 PCON:EQU 0FH
0022 ;
0023 ;
0024 ; ORG 3800H
0025 ;
0026 ENTER:MVI A,80H ;SET UP PORT
0027  OUT PCON ;
0028  MVI A,4 ;SET UP POSITION COUNTER
0029  STA PAD ;AND SAVE
0030  LXI H,TBL ;INITIALISE LOCATION POINTER
0031  SHLD VECT1
0032 ;
0033 ;NOTE THE FOREGOING IS ONLY USED ON FIRST
0034 ;ENTERING THE PROGRAM. EACH SUBSIQUENT CALL
0035 ;SHOULD BE ''RUN''.
0036 ;
0037 RUN:LHLD VECT1:POINT TO VECTOR?
0038  LDA PAD ;GET CURRENT SEQ. POSITION
0039  MOV C,A;SAVE IN STEP COUNTER
0040  LDA MOVE ;GET NUMBER OF STEPS
0041  MOV B,A ;
0042 RUN3:MOV A,M, ;GET PATTERN
0043  OUT APORT ;SEND TO MOTOR
0044  CALL DWELL ;WAIT FOR IT TO MOVE
0045  INX H ;POINT TO NEXT
0046  DCR B ;DOWN COUNTER
0047  JZ DONE ;FINISHED
0048  DCR C ;SEQUENCE DONE?
0049  JNZ RUN3 ;NO
0050  LXI H,TBL ;RESET POINTERS
0051  MVI C,4,
0052  JMP RUN3 ;AND CONTINUE
0053 ;
0054 DONE:SHLD VECT1 ;SAVE CURRENT POSITION FOR
0055  MOV A,C ;NEXT TIME
0056  STA PAD ;
0057  RET ;AND RETURN TO CALLING PROGRAM
0058 ;
0059 ;DWELL IS A SUBROUTINE WHICH CONTROLS
0060 ;THE SPEED OF THE MOTOR AND IS DEPENDANT
0061 ;UPON THE TYPE OF MOTOR AND ITS APPLICATION
0062 ;
0063 DWELL:PUSH H
0064  PUSH D
0065  LXI D,-1 ;SET UP A DECREMENTER
0066  LHLD TIME ;GET DELAY
0067 DW1 :DAD D; DEC. HL
0068  JC DW1 ;LOOP UNTIL HL=0
0069  POP D
```

*continued*

```
0070  POP H
0071  RET
0072  ;
0073  ;PROVIDE PROGRAM SCARTCH PAD
0074  ;
0075  PAD:DS 1 ;SAVE C-REG.
0076  VECT1:DS 2 ;SAVE HL
0077  ;
0078  ;SCRATCH PAD FOR CONSTANTS
0079  ;
0080  TIME:DS 2 ;DELAY FOR MOTOR STEPS
0081  MOVE:DS 1 ;HOLDS NUMBER OF STEPS
0062  ;
0083  ;TABLE OF STEPS
0084  ;
0085  TBL:DB 3,6,12,9
0086  ;
0087  END
```

Figure 2.43  *Power interfacing stepper motors*

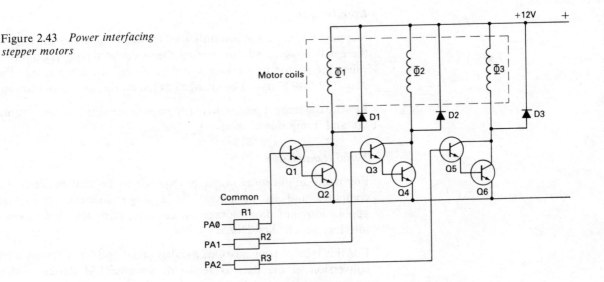

Stepper motors need power interfacing and Figure 2.43 shows an interface and connections to the 8255 for this application. It is important that the +12 V supply is isolated from the +5 V of the processor, and 5.6 V zener diodes may be connected between each peripheral pin and 0 V if desired. Diodes D1 to D4 are included to remove the back e.m.f. generated by switching the motor coil current from damaging the transistors, but this can reduce the available torque marginally.

An increase in torque can be achieved by using say a 48 V supply and fitting a series resistor of three times the coil resistance.

Assemble and enter the program. Enter the required number of degrees of rotation in memory location TURN: and test the program. Remember that decimal numbers must be converted to hex before entry, i.e. for one complete turn the entry will be $200_{(10)} = CB_{(16)}$.

*Question*

Select suitable values to provide 45° and 90° of rotation.

What error would have to be tolerated if a rotation of 50° was required?

*Answer*

50° rotation represents twenty-seven steps plus one half-step of $0.9° = 49.5° = -1.5°$ or twenty-eight steps of $1.8 = 50.4° = +0.4°$.

Stepper motors may be coupled to mechanical devices in numerous ways. Each can introduce some extra error into the system owing to backlash. A typical application of stepper drives is discussed in Chapter 5.

### Exercise 2.26

Stepper motors cannot instantly start their load moving from rest, so there may be some 'slip' at start-up. One method of reducing this is to cause the motor to increase speed gradually. In other words the period of 'dwell' has to be changed as the motor runs from start up.

Rewrite the stepper motor drive program to provide a smooth 'ramp up' and 'ramp down' of speed.

### Serial transmission

For many applications, data is required to be sent between the computer and its peripheral by a single connection. Typical applications are data storage on cassette, teletypes, and modem interface to telephone lines.

For this type of transmission, parallel–serial and/or serial–parallel conversion of the data is required. Suitable LSI devices which perform the required functions are known as:

1   Universal asynchronous receiver-transmitter (UART): typical type numbers ICM 6402 and AY5 1013.
2   Asynchronous communications interface adaptor (ACIA): from Motorola, type MC6850.
3   Programmable communications interface (PCI): from Intel, type 8251.
4   Serial input/output (SIO): from Zilog, type Z80 SIO.

If a synchronous formatting of the transmitted data is required then a universal synchronous–asynchronous receiver–transreceiver (USART) will need to be used. The PCI of Intel is a device of this type (see data sheet, Appendix 2).

*A serial interface*

The PCI type 8251 interfaces with the HEKTOR bus as shown in Figure 2.44.

Figure 2.44 *Serial I/O using the 8251*

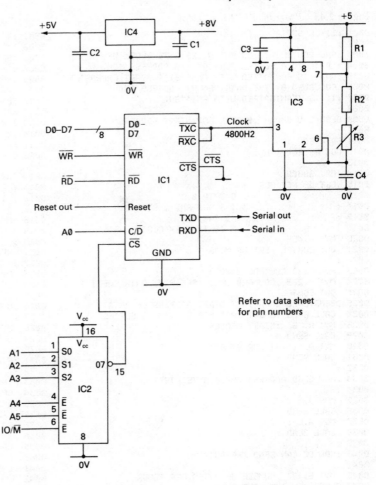

The data bus is connected directly to the PCI circuit as are the RD, WR and RESET lines.

Each PCI occupies two I/O addresses, one for control/status registers and the other for transmit and receive data. The register is selected by the $C/\overline{D}$ line which is supplied from A0.

The full address is determined by the address decoder IC2, in a similar manner to that of the parallel interface board. Here A1 to A5 are used to activate the decoder with the $IO/\overline{M}$ line. As the 0 output from the decoder is used this puts the I/O address at 00 and 01, but as A6 and A7 are not used in this application then the device will appear at addresses 40H, 80H, and C0H. This is of liitle consequence here as the I/O address map is not particularly crowded.

A ×16 serial (baud rate) clock is provided by IC3 (NE555) for the $\overline{T\times C}$ and $\overline{R\times C}$ inputs. This signal is at sixteen times the transmission rate, so for 300 bauds the astable must be set to 4800 Hz.

Figure 2.45   *Program SERIALCOMM*

```
0001 ;TITLE SERIAL OUT
0002 ;
0003 ;SUB-ROUTINE TO OUTPUT A 128 BYTE BLOCK OF
0004 ; DATA TO A SERIAL PORT AT I/O ADDRESS 10H
0005 ;DATA IS FORMATTED WITH AN ''S1'' START CHARACTER
0006 ;FOLLOWED BY THE DATA MEMORY ADDRESS AND
0007 ;IT IS TERMINATEED WITH AN ''S9''.
0008 ;
0009 BEGA:EQU 3000H ;BUFFER FOR START ADDRESS
0010 ;
0011 DATA:EQU 10H
0012 CONTRL:EQU 11H
0013 ;
0014  ORG 3800H
0015 ;SET UP PCI FOR :-
0016 ;                    8 DATA BITS
0017 ;                    1 START AND 1 STOP BIT
0018 ;                    NO PARITY
0019 ;                    X 16 TX & RX CLOCK
0020  MVI A,4EH ;
0021  OUT CONTRL ;SET UP MODE
0022 ;
0023 ;NOW SEND CONTROL WORD
0024  MVI A,27H ;COMMAND WORD (SEE P10, DATA SHEET.)
0025  OUT CONTRL ;
0026 DSEND:LHLD BEGA ;GET START ADDRESS OF DATA
0027  CALL PAWS ;SHORT WAIT FOR SYSTEM TO SETTLE
0028  MVI A,'S' ;START HEADER
0029  CALL SEND ;
0030  MVI A,'1':S1 FOR START
0031  CALL SEND ;
0032 ;
0033 ;NOW SEND ADDRESS UPPER BYTE FIRST
0034 ;
0035  MOV A,H ;
0036  CALL SEND
0037  MOV A,L ;
0038  CALL SEND
0039 ;
0040 ;NOW WE CAN SEND THE DATA
0041 ;
0042  MVI B,127 ;NUMBER OF BYTES PER BLOCK
0043 SLOOP:MOV A,M ;GET BYTE
0044 (ED THIS LINE MISSING!!!!)
0045 (THIS LINE SEEMS TO BE MISSING ALSO)
0046 (CANNOT READ THIS LINE)
0047  JNZ SLOOP
0048  RET ;RETURN TO CALLING PROGRAM
0049 SEND:PUSH PSW ;SAVE DATA
0050 LOOKU:IN CONTRL ;NOW SEE IF PCI READY
0051  RRC ;TX EMPTY BIT TO CY FLAG
0052  JNC LOOKU ;WAIT UNTIL READY
0053  POP PSW ;REGAIN DATA
0054  OUT DATA ;SEND DATA
0055  RET ;
0056 ;SHORT DELAY ROUTINE
0057 ;
0058 PAWS:PUSH H ;SAVE CURRENT
0059  PUSH D ;REGISTERS
0060  LXI H,8000H ;
0061  LXI D,-1 ;
0062 PAW: DAD D ;
0063  JC PAW
0064  POP D ;
0065  POP H;
0066  RET
0067 ;
0068  END
```

```
0001 ;TITLE - SIN
0002 ;
0003 ;SERIAL INPUT SUBROUTINE
0004 ;
0005 ;ROUTINE SEARCHES FOR AN ''S1'' SEQUENCE
0006 ;AFTER THIS IS FOUND THEN THE NEXT TWO
0007 ;BYTES ARE LOADED INTO THE HL REGISTER TO
0008 ;ACT AS A MEMORY POINTER FOR THE DESTINAT
0009 ;OF THE FOLLOWING DATA.
0010 ;
0011 DATA:EQU 10H
0012 CONTRL:EQU 11H
0013 ;
0014  ORG 3900H
0015 ;
0016 ;FIRST SET PCI FOR RECEIVE
0017 ;
0018  MVI A,40H;RESET PCI
0019  OUT CONTRL ;
0020 ;
0021 ;NOW PROGRAM IT, FIRST THE MODE
0022 ;
0023  MVI A,4EH ;
0024  OUT CONTRL ;
0025 ;
0026 ;NOW THE COMMAND WORD
0027 ;
0028  MVI A 06 ;
0029  OUT CONTRL ;
0030 ;
0031 ;FIND S1 SEQUENCE
0032 ;
0033 FINDS: CALL LOOK ;INPUT ROUTINE
0034  CPI 'S' ;IS IT AN ''S''
0035  JNZ FINDS ;NO
0036  CALL LOOK ;S IS O.K. NOW FIND A 1
0037  CPI '1' ;IS IT A 1?
0038  JNZ FINDS;NO SO START AGAIN
0039  CALL LOOK ;S1 OK SO NOW GET ADDRESS
0040  MOV L,A ;LOW BYTE
0041  CALL LOOK ;NEXT
0042  MOV H,A ;UPPER BYTE
0043  MVI B,127 ;NUMBER OF BYTES IN BLOCK
0044 LOAD:CALL LOOK ;GET DATA
0045  MOV M,A ;SAVE IT
0046  INX H
0047  DCR B
0048  JNZ LOAD ;
0049  RET ;ALL DONE
0050 ;
0051 ;INPUT ROUTINE
0052 ;
0053 LOOK: IN CONTRL ;
0054  RRC ;MOVE STATUS
0055  RRC ;BIT TO CY FLAG
0056  JNC LOOK ;
0057  IN DATA;CHARACTER RECEIVED SO GRAB IT
0058  RET ;AND RETURN WITH IT IN A-REG
0059 ;
0060  END
```

For this chip to transmit the $\overline{\text{CTS}}$ input must be low, otherwise the transmitter portion of the device is inhibited, so this pin (17) is connected to ground for the moment. Other modem control signals (RTS, DTR and DSR) will be ignored for the moment.

If you are unsure of the meaning of any of these abbreviations they can be found in the data sheet in Appendix 2.

The software driver for the serial interface is presented in Figure 2.45. It is two subroutines which may be called from a simple master program. The send program expects to find the start address of the data to be sent in a memory location having the label BEGA (address 3000H in the example). Data is formatted with an S1 to identify the start of the data block (why is identification necessary?), followed by the memory address of the data. Then a block of 128 bytes is sent.

It is usual to supply some form of error checking when transmitting data serially; this may be in the form of simple parity checking, or a check sum is sent at the end of each block. It is left to the student to implement and test these facilities.

The receive program searches the input for an S1 sequence; then the next two bytes are used as a memory pointer for data storage.

Generally more than 128 bytes will need to be transmitted, so your calling program will have to provide facilities for setting the start and end address of the data to be sent.

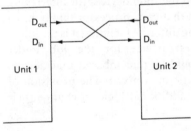

Figure 2.46 *Coupling two serial devices*

### Exercise 2.27

Set up the necessary hardware and load some suitable data into memory. Load the program of Figure 2.44. Set up start address and end address. Run the program. Using an oscilloscope, check the data output. (Better still, if you have a serial data analyser use this to check the output.)

### Exercise 2.28

Arrange to cross-couple the serial output of one serial board to the input of another (Figure 2.46) and check that data can be sent between them.

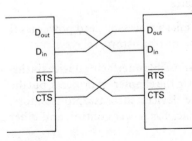

Figure 2.47 *Connecting the control signals*

### Exercise 2.29

Disconnect the $\overline{\text{CTS}}$ lines from ground and cross-couple it with the opposite $\overline{\text{RTS}}$ between the two units (Figure 2.47). Modify the program to set up the appropriate protocols for data transmission.

The appropriate commands are listed on page 10 of the data sheet.

When it is required to send data from one unit to the other, the sending unit writes a command word to the PCI in which bit D5 is high. This will enable the transmit section of the other PCI and any data pending to be sent until complete or the $\overline{RTS}$ is set high by putting bit 5 in the PCI command register low.

*Exercise 2.30*

Modify the above programs to provide check-sum error detection for the transmitted signal.

An easy way to work a check-sum is simply to add all the bytes together, discarding any carry, and then to negate (form two's complement) the sum and send this as a further byte in the block. When the receiver adds this to the sum of the received bytes, the total should be zero. Any other value indicates an error and appropriate action can be taken.

## 2.5  Conclusion

This chapter has concentrated on making the microprocessor 'do something' rather than just be a 'calculator'.

It has emphasized the fact that it is easy to use the device as a logic control system, but as there is a conflict in the requirements of the processor operation and outside process operation, some form of interface is necessary. This can range from a simple addressable latch as shown in the 8212 device to the advanced facilities of the 8255. Whilst the text has concentrated on the Intel devices for interfacing with HEKTOR, it should be noted that each processor family has a similar range of devices each offering differing facilities: it is the job of the technician to select the best device for the job under consideration. Additionally it should be remembered that devices from different manufacturers can be intermixed. The provision of suitable control signals may need additional logic to change their polarities or provide timing.

Whilst the methods presented here show one way of solving a problem, they are not the only answer. Sometimes it is easier and more desirable to provide a hardware solution; at other times a software one is the more appropriate.

Almost all devices external to the microprocessor need power level interfacing; this will involve the use of transistors, relays etc.

It has not been possible to present all possible interfacing devices in this chapter. Other devices will be found in the chapter on projects, but the reader is encouraged to investigate devices such as opto-isolators, optically coupled thyristors and triacs for power control, and other opto devices for data transmission.

One important thing to remember is that most things do not happen in zero time, so allow delays in programs for the action to take place.

# Chapter 3 **Programmable memory**

*Objectives of this chapter*    *When you have completed this chapter, you should be able to:*

*1   Select non-volatile programmable memory devices.*
*2   Understand the operating conditions for non-volatile programmable memory devices.*
*3   Use a PROM programmer to store an object code program in a programmable memory.*
*4   Know the procedure for erasing EPROMs.*

## 3.1   Introduction

Non-volatile progammable memory devices can be roughly divided into two main types – permanent (PROM) and erasable (EPROM). A further group – the electrically alterable read-only memory (EAROM), which can be considered semi-permanent – is also gaining popularity.

## 3.2   PROM

Permanent programmable read-only memory (PROM) is usually of the fusible link type; that is, the manufacturer produces the device with all memory cells set to a predetermined state and the program is entered by selectively blowing fusible links in the device. Figure 3.1 shows a typical output circuit for a PROM.

It is important to note that different manufacturers adopt different conventions when producing these devices. Some make the devices so that stated outputs are at logic 1, others so that the stated outputs are at logic 0.

Clearly it doesn't matter which convention is adopted, providing the programmer is aware of it.

With this type of chip the programming is permanent; almost no correction is possible. Once a bit is programmed it is not possible to return it to its original state, although unprogrammed bits can still be programmed.

When programming one of these devices (normally using the PROM programming facilities of a development system), it is important to follow the precise sequence specified by the manufacturers of the

Figure 3.1   *Output circuit of an*
*unprogrammed bipolar PROM (source:*
*National Semiconductor)*

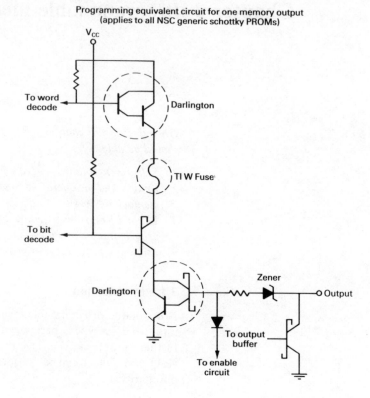

Programming equivalent circuit for one memory output
(applies to all NSC generic schottky PROMs)

equipment or an unreliable device may be produced. A typical
sequence might be:

1   Ensure that the temperature is within the manufacturers' limits
    for programming.
2   Select the address that is required to be programmed and allow
    the system to stabilize.
3   Increase $V_{cc}$ to programming level at a rate not exceeding that
    specified.
4   Connect the bit to be programmed to the correct supply (0 V or
    $-V_{cc}$).
5   Enable the chip for programming time (about 10 μs).
6   Reduce voltages to normal and verify programming has occurred.
7   Repeat 3–6 five more times to ensure programming.
8   Repeat steps 2–7 for each other bit to be programmed.

## 3.3   The EPROM

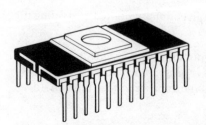

Figure 3.2   *The appearance of an*
*EPROM*

For system development and small-scale production, the erasable
PROM is generally used. Erasable in this case usually means

ultraviolet erasable, that is, the contents of the memory can be removed by exposing the chip to shortwave u.v. light, a window is provided in the top of the device for the u.v. to enter. This window makes such devices easy to recognize (Figure 3.2).

Table 3.1 gives some typical types and sizes of EPROMs which are, at present, in general use.

**Table 3.1** *EPROMs currently in use*

| Type no. | Size (bytes) | Notes |
|---|---|---|
| 1702A | 512 × 8 bits | Now almost obsolete |
| 2708 | 1024 (1 K) × 8 | 3-level supply |
| 2508 | 1024 (1 K) × 8 | Single supply |
| 8741/48 | 1024 × 8 | Single-chip CPU with EPROM |
| 2716 | 2 K × 8 bits | Single supply |
| 2516 | 2 K × 8 bits | Single program cycle |
| 8755 | 2 K × 8 bits | ROM I/O chip. |
| 2732 | 4 K × 8 bits | |
| 2532 | 4 K × 8 bits | |
| 2764 | 8 K × 8 bits | |

*Note*: 25XX and 27XX devices are not totally pin compatible.

### Programming requirements for EPROMs

The programming requirements vary slightly with different types of EPROM but, in general, the most popular types of 2716/2732/2764 follow a similar line.

In this section we examine the hardware and software programming requirements of a 2716.

In the erased state all bits are set to 1 and the device is programmed by setting 0s into the correct places. This is done by:

1  Setting the $V_{pp}$ pin to +25 V.
2  Setting the $\overline{CS}$ pin at logic 1 (+5 V).
3  Apply data BYTE to the output pins.
4  Apply address of byte to be programmed and allow to stabilize.
5  Apply a 50 ms active high program pulse to the PD/PGM pin.
6  Repeat 1–5 for each location to be programmed.

It is possible to program any location individually, but once a bit has been programmed to a 0 it is only possible to return it to a 1 by exposure to u.v. light, and this will erase the whole device.

The sequence of events is summarized in the flow chart in Figure 3.3.

A hardware interface for HEKTOR is shown in Figure 3.4.

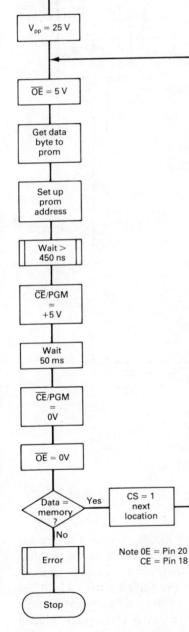

Figure 3.3  *EPROM programmer flow chart*

Figure 3.4   *EPROM programmer*
*circuit diagram*

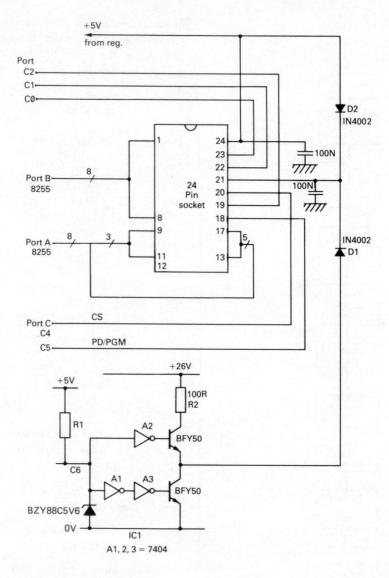

## 3.4   Erasing PROMS

The erasure characteristics of most PROMS is such that erasure
begins to occur when exposed to light with wavelengths shorter than
approximately 4000 ångströms. Sunlight and certain types of
fluorescent lamps have wavelengths in the range $3 \times 10^{-7} - 4 \times 10^{-7}$ m.
Data shows that constant exposure to room-level fluorescent lighting
could erase the typical 2716 in approximately three years, while it
would take approximately 1 week to cause erasure when exposed to
direct sunlight.

If an EPROM is to be exposed to this type of lighting then the window in the device should be covered with opaque tape.

### Erasure procedure

The recommended procedure for erasure is to illuminate the window of the EPROM with a u.v. source which has a wavelength of $2.537 \times 10^{-7}$ m (2537 ångströms). Most data sheets specify a distance of 1 inch from the lamp. The length of time for erasure is defined in terms of required energy incident upon the window and is expressed in watt-seconds per square centimetre ($Ws/cm^2$).

According to lamp manufacturers the output will decrease with age, leading to an increase in exposure time. The output of the lamp used for erasure should be checked periodically using methods specified by the manufacturers.

The sensors used with most u.v. intensity meters show reduced output with constant exposure to u.v. light. Therefore they should not be permanently placed inside the erasure enclosure; they should be used only for periodic measurements.

*It is essential that no u.v. lamp is operated in the open as the ultraviolet output from these tubes can cause eye damage.*

When operating an EPROM eraser, ensure that all interlock switches are operational and cut off the tube when the loading door is opened.

## 3.5  Programming EPROMs

The EPROM is one of the most widely used non-volatile memory devices, and many units are marketed for the programming of these devices. The methods employed can be roughly divided into manual and automated methods.

There are many applications where the manual method is of value and may be the only one available, whereas the automated method can be useful when the whole design process for a unit is totally computer controlled.

Some development systems, such as the one described in Chapter 1, incorporate a PROM programmer; others do not have this facility.

Stand-alone units have a buffer memory – 1 or 2 Kbytes – which holds the data intended for the EPROM. This can be loaded either from a small hexadecimal keyboard or from an interface from a magnetic or paper tape-reader or teletype.

Once the buffer is loaded then the software of the programmer handles the transfer between buffer and PROM. Some units have a

Figure 3.5 *The PRO-LOG PROM programmer*

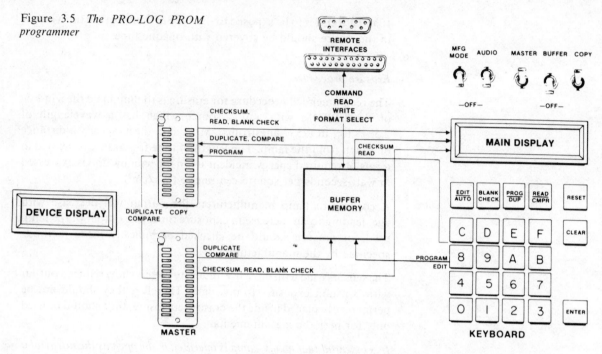

number of sockets so that, say, eight or twelve PROMs can be programmed at the same time.

Figure 3.5 shows the arrangement for a PRO-LOG M900 PROM programmer which has a range of plug-in 'personality modules' that allow a range of IC's to be programmed.

### An EPROM program for HEKTOR

This unit connects to the 8255 PPI board introduced in Chapter 2. It provides the minimum hardware for programming a device of the 2716 type. Note that only the 2716 devices listed in Table 3.2 are suitable without modifications to the unitor PPI board.

**Table 3.2** *2716 devices*

| Manufacturer | Type no. |
| --- | --- |
| Fairchild | F2716 |
| Intel | I2716–2 |
|  | 2716–1 |
| Mostek | MK2716 |
| Motorola | MCM2716 |
|  | MCM27A16 |
| NEC | μPD2716 |
| Signetics | 2716 |
| Texas Instrument | TM2516 |

Other devices can easily be accommodated by rearranging the connections around pins 18, 19, 20 and 21.

2732-type devices can be accommodated by using PC3 to supply address line to printer.

Port B and bits 0–2 of port C supply the EPROM address. Port A supplies the data byte and the rest of port C supplies the control signals.

PC6 is the programming voltage on/off bit. When the programming voltage is off the $V_{pp}$ pin is supplied with +5 V from the $V_{cc}$ supply.

D2 prevents the +26 V from reaching the $V_{cc}$ supply.

### The software interface

Owing to the limited memory on the standard HEKTOR system, the buffer for programming the EPROM will be located between 3400 and 3FFF, i.e. a 2 Kbyte buffer. The RAM version of the software driver will start at 3000. The stack will be located at 2FFF and down.

A number of address pointers need to be generated, i.e.:

1  Source buffer start address.
2  ROM START address.
3  ROM END address.

The program then calculates the number of bytes to be programmed.

The program is arranged to be interactive, with information being requested by the system as required. It is necessary for the object code intended for the EPROM to be resident in the buffer before this program is run. This can be arranged by use of the monitor LOAD command. Figure 3.3 is the outline flow chart for the program and Figure 3.6 is the listing of the program.

Figure 3.6 *PROM programmer software driver listing*

```
'Tektronix  8080/8085 ASM VD.3   PRMBL080

00002                    ;
00003                    ;           SOFTWARE DRIVER FOR HEKTOR FROM PROGRAMMER
00004                    ;
00005                    ;           DATA FOR TRANSFER SHOULD BE RESIDENT
00006                    ;           BETWEEN ADDRESSS 3000H AND 37FFH BEFORE
00007                    ;           THIS PROGRAM IS RUN
00008                    ;
00009                    ;           THIS PROGRAM HAS AN ORGIN OF 3800H AND
00010                    ;           IS INTERACTIVE, I.E. IT REQUESTS ANY DATA
00011                    ;           IT REQUIRES FROM THE OPERATOR
00012                    ;
00013                    ;           THE PROGRAM IS COMPATABLE WITH MOST
00014                    ;           2716-TYPE DEVICES.
00015                    ;
00016                    ;           TABLE OF EQUATES AND CONSTANTS
00017                    ;           ================================
00018                    ;
00019                    ;           I/O ADDRESSES                        continued
```

```
00020        000C        PORTA   EQU     0CH
00021        000D        PORTB,  EQU     0DH
00022        000E        PORTC   EQU     0EH
00023        000F        PCON    EQU     0FH
00024                    ;
00025                    ;       MONITOR CALLS
00026                    ;
00027        0057        MONW    EQU     57H             ;RESTART POINT
00028        02DA        CRLF    EQU     2DAH            ;PRINT NEW LINE
00029        02E7        PRSPC   EQU     2E7H            ;PRINT A SPACE
00030        030A        PRMES   EQU     30AH            ;PRINT A STRING
00031        0351        PRWD    EQU     351H            ;PRINT ADDRESS IN HL
00032        05BE        KYIN    EQU     5BEH            ;SINGLE KEY INPUT
00033        0660        KLUC    EQU     660H            ;CONVERT KEYCASE
00034        06C0        VDU     EQU     6C0H            ;SCREEN OUTPUT
00035                    ;
00036                    ;       CONSTANTS
00037                    ;
00038        1000        PULTIM  EQU     1000H           ;TIME CONSTANT FOR PULSE
00039        0040        READ    EQU     40H             ;PATTERNS FOR
00040        0030        PRG     EQU     30H             ;CHIP ENABLE
00041        0060        DSEL    EQU     60H             ;
00042                    ;
00043                    ;       CREATE BUFFERS
00044        3C00    >           ORG     3C00H           ;
00045 3C00   0002        DATST   BLOCK   2               ;DATA START ADDRESS
00046 3C02   0002        DATEND  BLOCK   2               ;DATA END ADDRESS
00047 3C04   0002        PRMST   BLOCK   2               ;PROM START ADDRESS
00048                    ;
00049                    ;
00050                    ;       PROGRAM START
00051                    ;
00052        3800    >           ORG     3800H
00053                    ;
00054 3800   CDDA02      GETAD   CALL    CRLF            ;NEW LINE
00055 3803   217F39  >           LXI H,  MSG1            ;MESSAGE STRING
00056 3806   CD0A03              CALL    PRMES           ;PRINT IT
00057 3809   219030              LXI H,  3000H           ;DEFAULT START ADDRESS
00058 380C   CD6D39  >           CALL    KEYIN           ;GET ANSWER TO QUESTION
00059 380F   FE59                CPI     ''Y''           ;ADDRESS O.K.?
00060 3811   C41E39  >           CNZ     NEWAD           ;GET NEW ADDRESS
00061 3814   22003C  >           SHLD    DATST           ;SAVE NEW START
00062 3817   CDDA02              CALL    CRLF            ;NEW LINE
00063 381A   21A039  >           LXI H,  MSG2            ;SECOND PROMPT
00064 381D   CD0A03              CALL    PRMES           ;PRINT IT
00065 3820   21FF37              LXI H,  37FFH           ;DEFAULT END ADDRES
00066 3823   CD6D39  >           CALL    KEYIN           ;GET ANSWER
00067 3826   FE59                CPI     ''Y''           ;ADDRESS O.K.?
00068 3828   C41E39  >           CNZ     NEWAD           ;
00069 382B   22023C  >           SHLD    DATEND          ;SAVE ADDRESS
00070 382E   21BF39  >           LXI H,  MSG3            ;NEW CHECK PROM ADDRESSES
00071 3831   CD0A03              CALL    PRMES           ;
00072 3834   210000              LXI H,  0               ;DEFAULT PROM START=0
00073 3837   CD6D39  >           CALL    KEYIN           ;GET CONFIRMATION
00074 383A   FE59                CPI     ''Y''           ;=YES?
00075 383C   C41E39  >           CNZ     NEWAD           ;NO
00076 383F   22043C  >           SHLD    PRMST           ;
00077                    ;
00078                    ;       NOW CALCULATE THE NUMBER OF BYTES
00079                    ;       TO BE PROGRAMMED
00080                    ;
00081 3842   2A003C  > NUMBYT    LHLD    DATST           ;GET LOW ADDRESS
00082 3845   EB                  XCHG                    ;HL -- DE
00083 3846   2A023C  >           LHLD    DATEND          ;END ADDRESS
00084                    ;
00085                    ;       NOW SUBTRACT HL AND DE
00086                    ;
00087 3849   CD7439  >           CALL    NEGDE           ;2 S COMPLIMENT DE
00088 384C   19                  DAD     D               ;HL-DE
```

*continued*

```
00089 384D  CD7439   >      CALL    NEGDE         ;RETURN DE TO POSITIVE
00090 3850  EB              XCHG                  ;AGAIN
00091                 ;
00092                 ; NOW HL=START ADDRESS OF DATA AND DE=NUMBER OF
00093                 ; BYTES TO BE PROGRAMMED
00094                 ;
00095                 ;      NEW GET PROM START ADDRESS IN BC
00096                 ;
00097 3851  3A043C   >      LDA     PRMST         ;
00098 3854  4F              MOV     C;   A        ;LOW BYTE IN C
00099 3855  3A053C   >      LDA     PRMST+1       ;
00100 3858  47              MOV     B,   A        ;HIGH BYTE IN B
00101                 ;
00102                 ; **************************************************
00103                 ;
00104                 ; NOW REGISTERS = BC...START ADDRESS FOR DATA TO PROM
00105                 ;                 DE...NUMBER OF BYTES TO BE PROGRAMMED
00106                 ;                 HL...START ADDRESS OF DATA BUFFER
00107                 ;
00108                 ; **************************************************
00109                 ;
00110                 ;      NOW SET UP THE PPI TO CHECK IF PROM ERASED
00111                 ;
00112 3859  3E90            MVI A, 90H            ;PORT A=IN, B & C=OUT
00113 385B  D30F            OUT     PCON          ;
00114                 ;      CHECK PROM
00115                 ;
00116                 ;
00117 385D  C5              PUSH    B             ;SAVE STATUS
00118 385E  D5              PUSH    D             ;
00119 385F  79       LOOK   MOV     A,C           ;SET UP PROM ADDRESS
00120 3860  D30D            OUT     PORTB         ;
00121 3862  78              MOV     A,B           ;UPPER BYTE
00122 3863  E60F            ANI     0FH           ;CLEAR UPPER NYBBLE
00123 3865  C640            ADI     READ          ;SET UP ENABLES ETC
00124 3867  D30E            OUT     PORTC         ;
00125 3869  DB0C            IN      PORTA         ;GET DATA
00126 386B  FEFF            CPI     -1            ;IS IT FFH
00127 386D  C2EB38   >      JNZ     NTCLN         ;
00128 3870  03              INX     B             ;O.K. SO FAR - MOVE POINTERS
00129 3871  15              DCR     D             ;AND COUNTERS
00130 3872  7A              MOV     A,D           ;CHECK FOR ALL DONE
00131 3873  B1              ORA     C             ;
00132 3874  C25F38   >      JNZ     LOOK          ;
00133 3877  3E60            MVI     A,DSEL        ;DESELECT PROM
00134 3879  D30E            OUT     PORTC         ;
00135 387B  D1              POP     D             ;RESTORE REGISTERS
00136 387C  C1              POP     B             ;
00137 387D  3E80     PRGRM  MVI     A,80H         ;RESET PORT FOR PROGRAMMING
00138 387F  D30F            OUT     PCON          ;
00139 3881  7E       PRGRM2 MOV     A,M           ;GET DATA TO BE PROGRAMMED
00140 3882  D30C            OUT     PORTA         ;TO PROM
00141 3884  79              MOV     A,C           ;ADDRESS IN PROM FOR DATA
00142 3885  D30D            OUT     PORTB         ;
00143 3887  78              MOV     A,B           ;GET UPPER ADDRESS
00144 3888  E60F            ANI     0FH           ;CLEAR ANY RUBBISH
00145 388A  C630            ADI     PRG           ;SET POWER AND ENABLES ETC.
00146 388C  D30E            OUT     PORTC         ;TO PORT
00147 388E  CD9438   >      CALL    PUSLE         ;PROGRAM PULSE
00148 3891  3E90            MVI     A,90H         ;NOW CHECK FOR PROGRAMMING
00149 3893  D30F            OUT     PCON          ;SET A FOR INPUT
00150                 ;
00151 3895  3E08            MVI     A,8           ;SET UP ENABLES ETC.
00152 3897  D30F            OUT     PCON          ;
00153 3899  3E09            MVI     A,9           ;
00154 389B  D30F            OUT     PCON          ;SET BITS ON PORT C
00155 389D  DB0C            IN      PORTA         ;GET DATA FROM PROM
00156 389F  F40600          CP      M             ;PROM=MEMORY
00157 38A2  C2CB38   >      JNZ     PRGERR        ;NO, ERROR
```

*continued*

```
00158 38A5  3E0A          MVI   A,10        ;DESELECT CHIP
00159 38A7  D30F          OUT   PCON        ;
00160 38A9  23            INX   H           ;NOW TO THE NEXT
00161 38AA  03            INX   B           ;
00162 38AB  15            DCR   D           ;AND SEE IF ALL DONE
00163 38AC  7B            MOV   A,E         ;
00164 38AD  B2            ORA   D           ;
00165 38AE  C28138   >    JNZ   PRGRM2      ;CONTINUE
00166 38B1  C35700        JMP   MONW        ;ALL DONE
00167                 ;
00168                 ;
00169 38B4  3E0B    PULSE MVI   A,11        ;SET UP POWER
00170 38B6  D30F          OUT   PCON        ;
00171 38B8  D5            PUSH D            ;SAVE STATUS
00172 38B9  E5            PUSH H            ;
00173 38BA  210010        LXI   H,PULTIM    ;TIME CONSTANT FOR PULSE
00174 38BD  11FFFF        LXI   D,-1        ;
00175 38C0  19      LOOPN DAD   D           ;
00176 38C1  DAC038        JC    LOOPN       ;DELAY LOOP
00177 38C4  E1            POP   H           ;
00178 38C5  D1            POP   D           ;RESTORE
00179 38C6  3E0A          MVI   A,10        ;
00180 38C8  D30F          OUT   PCON        ;PULSE RESET
00181 38CA  C9            RET               ;AND RETURN
00182                 ;
00183                 ;
00184???????????
00185 38CB  E5      PRGERR PUSH H           ;
00186 38CC  21D938   >    LXI   H,EMSG      ;
00187 38CF  CD0A03        CALL  PRMES       ;
00188 38D2  E1            POP   H           ;RESTORE ADDRESS
00189 38D3  CD5103        CALL  PRWD        ;PRINT ADDRESS
00190 38D6  C35700        JMP   MONW        ;RETURN TO MON1TOR
00191                 ;
00192 38D9  50524F47 EMSG ASCII ''PROGRAM ERROR AT''
00192 38DD  52414D20
00192 38E1  4552524F
00192 38E5  52204154
00192 38E9  20
00193 38EA  00            BYTE  0
00194                 ;
00195                 ;
00196 38EB  21FC38  >NTCLN LXI   H,DMSG      ;
00197 38EE  CD0A03        CALL  PRMES       ;
00198 38F1  CD6D39   >    CALL  KEYIN       ;GET ANSWER
00199 38F4  FE59          CPI   ''Y''       ;ANSWER = YES
00200 38F6  C25700        JNZ   MONW        ;NO
00201 38F9  C37D38        JMP   PRGRM       ;PROGRAM ANYWAY
00202?               ;
00203 38FC  50524F4D DMSG ASCII ''PROM NOT ERASED, PROGRAM ANYWAY?''
00203 3900  204E4F54
00203 3904  20455241
00203 3908  5345442C
00203 390C  2050524F
00203 3910  4752414D
00203 3914  20414E59
00203 3918  5741593F
00203 391C  20
00204 391D  00            BYTE  0
00205                 ;
00206 391E  CDDA02  NEWAD CALL  CRLF        ;
00207 3921  21DF39   >    LXI   H,MSG4      PROMPT MESSAGE
00208 3924  CD0A03        CALL  PRMES
00209 3927  CD3039   >    CALL  INHEX       ;GET NEW ADDRESS
00210 392A  67            MOV   H,A         ;AND PUT IN HL
00211 392B  CD3039   >    CALL  INHEX       ;
00212 392E  6F            MOV L, A          ;LOWER BYTE
00213 392F  C9            RET
00214                 ;
```

*continued*

```
00215 3930  CD6D39  > INHEX  CALL  KEYIN      ;GET KEYBOARD
00216 3933  CDC006           CALL  VDU        ;ECHO INPUT
00217 3936  CD4B39  >        CALL  CONV       ;ASCII --> HEX
00218 3939  07               RLC              ;POSITION NYBBLE
00219 393A  07               RLC
00220 393B  07               RLC
00221 393C  07               RLC
00222 393D  47               MOV   B,A        ;SAVE DATA
00223 393E  CD6D39  >        CALL  KEYIN      ;
00224 3941  CDC006           CALL  VDU        ;
00225 3944  CD4B39  >        CALL  CONV       ;
00226 3947  E60F             ANI   0FH        ;CLEAR UPPER NIBBLE
00227 3949  80               ADD   B          ;FORM BYTE
00228 394A  C9               RET
00229 394B  D630     CONV    SUI   30H        ;KEY - 48
00230 394D  FA6239  >        JM    NOTX       ;NOT HEX
00231 3950  FE0A             CPI   0AH        ;NUMBER  10?
00232 3952  DA6139  >        JC    GOTIT      ;NO 0 - 9
00233 3955  FE11             CPI   11H        ; 16?
00234 3957  DA6239  >        JC    NOTX       ;TOO BIG
00235 395A  FE17             CPI   17H
00236 395C  D26239  >        JNC   NOTX       ;
00237 395F  D607             SUI   7          ;NUMBER A - F
00238 3961  C9       GOTIT   RET              ;
00239 3962  3E3F     NOTX    MVI   A,3FH      ;ASCII ?
00240 3964  CDC006           CALL  VDU        ;PRINT IT
00241 3967  CD6D39  >        CALL  KEYIN      ;GET CORRECTION
00242 396A  C34B39  >        JMP   CONV       ;
00243                 ;
00244                 ;
00245 396D  CDBE05   KEYIN   CALL  KYIN       ;GET KEYBOARD
00246 3970  CD6006           CALL  KLUC       ;CONVERT TO LOWER CASE
00247 3973  C9               RET              ;
00248                 ;
00249                 ;   FORM 2-S COMPLIMENT IN DE
00250                 ;
00251 3974  B7       NEDGE   ORA   A          ;CLEAR CARRY
00252 3975  7B               MOV   A,E        ;GET LO BYTE
00253 3976  2F               CMA              ;COMPLEMENT (ONES)
00254 3977  3C               INR   A          ;AND INC FOR TWOS
00255 3978  5F               MOV   E,A        ;AND RETURN TO REGISTER
00256 3979  7A               MOV   A,D        ;GET UPPER BYTE
00257 397A  2F               CMA              ;
00258 397B  C600             ADI   0          ;ADD IN CARRY FOR LOWER BYTE
00259 397D  57               MOV   D,A        ;
00260 397E  C9               RET              ;
00261                 ;
00262 397F  44415441 MSG1    ASCII ''DATA START ADDRESS 3000H, O.K.?''
00262 3983  20535441
00262 3987  52542041
00262 398B  44445245
00262 398F  53532033
00262 3993  30303048
00262 3997  2C204F2E
00262 399B  4B2E3F20
00263 399F  00               BYTE  0
00264 39A0  44415441 MSG2    ASCII --DATA END ADDRESS 37FFH, O.K.?''
00264 39A4  20454E44
00264 39A8  20414444
00264 39AC  52455353
00264 39B0  20333746
00264 39B4  46482C20
00264 39B8  4F2E4B2E
00264 39BC  3F20
00265 39BE  00               BYTE  0
00266 39BF  50524F4D MSG3    ASCII --PROM START ADDRESS 0000, O.K.?''
00266 39C3  20535441
00266 39C7  52542041
00266 39CB  44445245
```

*continued*

```
00266 39CF  53532030
00266 39D3  3030302C
00266 39D7  204F2E4B
00266 39DB  2E3F20
00267 39DE  00              BYTE    0
00268 39DF  4E455720    MSG4  ASCII ‐‐NEW ADDRESS ?‘‘
00268 39E3  41444452
00268 39E7  45535320
00268 39EB  3F20
00269 39ED  00              BYTE 0
00270                   ;
00271                       END
```

Tektronix 8080/8085 ASM V3.3 Symbol Table

scalars

```
A ----- 0007 B ----- 0000 C ----- 0001 CRLF -- 02DA D ----- 0002
DSEL -- 0060 E ----- 0003 H ----- 0004 KLUC -- 0660 KYIN -- 05BE
L ----- 0005 A ----- 0006 MONW -- 0057 PCON -- 000F PORTA - 000C
PORTB - 000D PORTC - 000E PRG --- 0030 PRMES - 030A PRSPC - 02E7
PRWD -- 0351 PSW --- 0006 PULTIM 1000 READ -- 0040 SP ---- 0006
VDU --- 06C0
```

% (default) Section (3C06)

```
CONV -- 394B DATEND  3C02 DATST - 3C00 DMSG -- 38FC EMSG -- 38D9
GETAD - 3800 GOTIT - 3961 INHEX - 3930 KEYIN - 396D LOOK -- 385F
LOOPN - 38C0 MSG1 -- 397F MSG2 -- 39A0 MSG3 -- 39BF MSG4 -- 39DF
NEGDE - 3974 NEWAD - 391E NOTX -- 3962 NTCLN - 38EB NUMBYT 3842
PRGERR  38CB PRGRM - 387D PRGRM2  3881 PRMST - 3C04 PUSLE - 38B4
```

271 Source Lines  271 Assembled Lines  47272 Bytes available

          No assembly errors detected

# Chapter 4  Fault-finding microprocessor systems

*Objectives of this chapter*  *When you have completed this chapter, you should be able to:*

1  *Identify typical faults which can occur in a microprocessor-based system.*
2  *Explain the limitation of conventional test gear in microprocessor-based system fault-finding.*
3  *Apply to a microprocessor-based system the following diagnostic aids:*
   *(a)  Logic probes*
   *(b)  Signature analyser*
   *(c)  Data/logic analyser*
   *(d)  Software diagnostics.*

## 4.1  Introduction

A good microprocessor-based system will have fault-finding aids built in from the design stage. This provision can be classified under the general heading of built-in test equipment (BITE) and can be of considerable help in locating the stage which is malfunctioning.

The approach to fault-finding will depend on the environment and, for the purpose of this chapter, it is assumed that the usual inspection for obvious short-circuits due to solder splashes and open-circuits due to bad printed circuit boards or dry joints has been implemented and all such obvious faults rectified.

Fault-finding in microprocessor-based systems is, in general, little different to that in other digital logic circuits. As with any circuit, a good knowledge of the system is necessary. A study of the theory of operation, the block diagrams and circuit diagrams provides a basis from which to work.

A number of problems are peculiar to fault-finding in micro-processor-based systems because in such systems:

1  Control is by software.
2  The system cannot be stopped for examination.
3  The system operates very quickly.
4  Bus structured connections are used.

A number of specialized instruments have been developed for use with these systems in addition to the familiar oscilloscope.

Five points are particularly worth making:

1  Microprocessors are sequential machines. Program flow depends on a sequence of instructions. If a single bit in an instruction or data byte is in error then the whole system may crash. Noise, glitches and bad memory bits are the most common sources of single-bit errors. These failures are difficult to pinpoint because the whole system may operate incorrectly. Other sources of single-bit errors will be examined later.

2  Most CPUs are dynamic devices; hence although the external circuit may be held in a fixed state, the processor (and dynamic RAM) must be kept running. It is usual to employ fault-finding methods which are dynamic and operate in system time.

3  The signals on the processor bus can be at a speed in excess of 4 MHz, which puts severe constraints on the test gear if meaningful results are to be obtained and/or the system is not disturbed by the application of the measuring equipment. The system cannot easily be slowed down as dynamic cells in the CPU need frequent refreshment. If this is inadequate, data will be lost.

4  Bus structures make it possible to connect many complex devices to a common line. Finding one bad device can present problems but the current probe/pulser can help solve this problem.

5  The data bus acts as a digital feedback path for bad signals. This will propagate the errors through good chips and back to the source. The best way to deal with this problem is to open the feedback loops somewhere.

## 4.2  Some basic faults

### System clocks

Poor clock signals can cause a number of apparent faults from non-operation to semi-functional activity and frequent program crashes.

Clock voltages are generally specified wider than standard TTL waveforms and usually need the full supply voltage swing. The rise and fall times must be within the specification of the processor.

Crystals can sometimes break into an overtone mode of operation, which gives a clock which is too fast for the system. Alternatively, if the clock is too slow, dynamic register cells in the CPU will not be refreshed quickly enough and data will be lost.

### Power-on restart

The reset signal to the CPU can cause problems. Generally the reset line must be active for a set minimum number of clock pulses; a reset pulse which is too short, noisy or too slow in transition can start the system off at the wrong point. This may lead either to erratic activity or no activity at all.

### Signal degradation

The long parallel bus and control lines used in some systems are sometimes susceptible to cross-talk and transmission line effects (reflections etc.). These problems can show up as glitches on adjacent lines or ringing (multiple transitions) or spurious signals on the driven line. Any of these conditions can cause faulty data which is difficult to detect. These problems are most severe when lines are long and the system is running at the limits of its speed and driving power. They will be aggravated by the use of extender cards.

## 4.3 Simple measurements

We will start with some simple measurements.

The circuit diagram of HEKTOR is shown in Figure 4.1. One of the most important things to check first is that the power supply to all the ICs is correct. Sometimes circuits do not show the power connections to the chips, so it will be necessary to look up the power connections in a data book or on a data sheet.

### Exercise 4.1

1   Using a conventional voltmeter (one with an impedance greater than 20 kΩ/V will be needed, with data sheets), identify the power pins of each IC in HEKTOR and check the voltage levels.
2   Repeat (1) using a CRO.
  (a)   Which is the more reliable method?
  (b)   What are the voltage limits for proper operation of the ICs?

The 5 V power lines must be between the limits 4.75 V and 5.25 V otherwise logic levels cannot be guaranteed.

### Exercise 4.2

Using the CRO and probe examine the clock signals on the CPU and peripheral chips. Compare these with the requirements specified in the device data sheets.

1   What are the voltage levels of the clock waveform?
2   What is the clock frequency?
3   Now look at the signals occurring on the address/data bus (pins 12 to 19). Can these signals be resolved?
4   What effect is the CRO probe having on the signals?

Figure 4.1  *Circuit diagram of*
*HEKTOR*

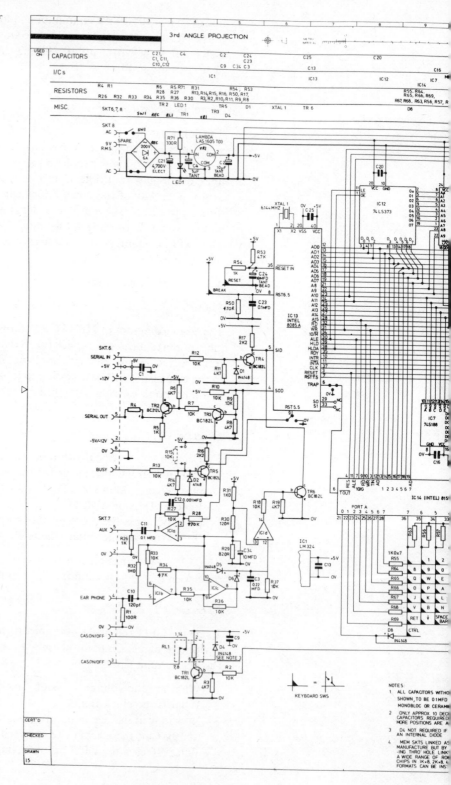

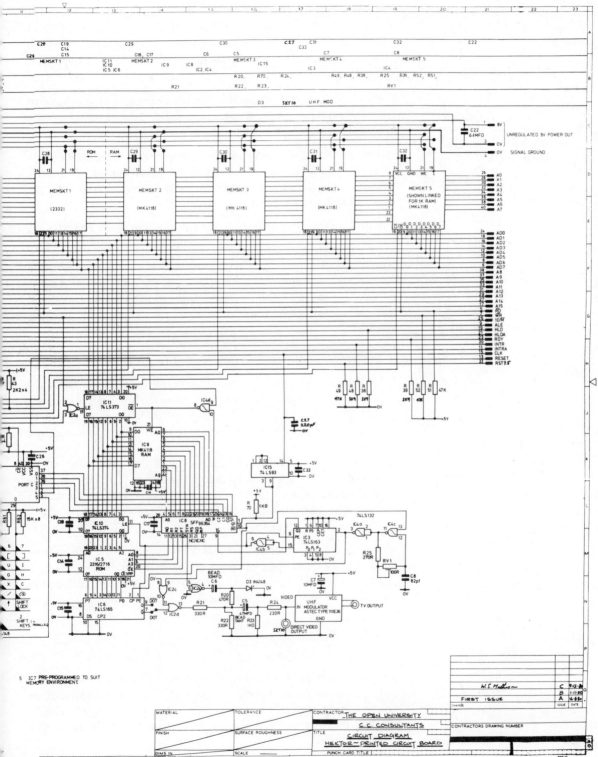

5  IC7 PRE-PROGRAMMED TO SUIT
   MEMORY ENVIRONMENT

CONTRACTOR  THE OPEN UNIVERSITY
            C.C. CONSULTANTS
TITLE  CIRCUIT DIAGRAM
       HEKTOR—PRINTED CIRCUIT BOARD

FIRST ISSUE

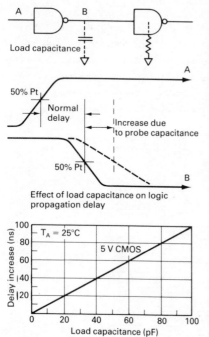

Effect of load capacitance on logic
propagation delay

Figure 4.2   *Illustration of probe
loading*

The X1(1) and X2(2) signals are almost a triangular wave, this is the oscillator circuit and should be at a frequency of 6.14 MHz in HEKTOR. Notice how this is cleaned up when it appears on the CLK(OUT) pin (37) of the central processing unit at a frequency of 3.12 MHz.

The address/data bus signals are somewhat of a mess, to say the least! This illustrates one problem of bus-orientated devices. What is really needed is something to select just parts of the waveform.

Errors can be introduced by the measuring device. If the oscilloscope has a narrow bandwidth, distortion of the waveform can occur. Rise time can be an important factor with some devices and its measurement can be a problem.

Figure 4.2 illustrates how the CRO probe and leads can affect a waveform.

To minimize the probe's effect on the circuit, the smallest possible load should be imposed, both d.c. load and, more importantly, reactively. Loading which is due largely to cable capacitance will produce severe degradation of the signals in any high-speed switching circuit.

A number of solutions are available to this problem:

1   10X *passive probes*   These are commonly available and have the added advantage of ruggedness, long reach and high impedance (10 MΩ and approximately 10–20 pF). However, they can be awkward if more than two or three need to be connected to a circuit.
2   *Active probes*   These combine many of the advantages of the passive probes with even greater performance. They are available with, typically, bandwidths up to 500 MHz and input impedances over 100 MΩ. However, they can be somewhat bulky and are definitely less rugged.

As was seen when examining the data bus signals, little information can be gained from the CRO trace except that signals are present. A number of special-purpose test instruments have been developed to ease the problem. Among these are:

1   Signature analysers.
2   Logic analysers.
3   In-circuit emulators.

Repair or testing of a microprocessor-based system usually starts with getting the system *kernel* operational. The kernel consists of the power supply system, clock and CPU. The power supply can be checked with conventional test gear, but remember the limits for TTL circuit supplies.

When the clock is running the address bus can next be checked for activity and hence the control bus and address decoders etc. When these are operational the data bus can be checked for activity. If a more detailed examination of bus activity is required then a data or logic analyser will be necessary.

Provided the CPU and its immediate ROM and RAM are operational, then the remainder of the system can be checked with diagnostic software of a type similar to that which is used during a self-test program.

The test programs presented here are developed for the HEKTOR unit, but are transportable to other processor systems of a similar type. Single-chip processor systems present special problems and will not be considered here.

## 4.4 Memory system checks

Memory failure in a microprocessor system from a total system failure to a single bit of faulty data can produce deviant system failure.

Most memory failures can be found during the power-up self-test program, unless the memory failure prevents this program from running. RAM failures occurring in the area of the memory used for the stack will usually cause the system to crash, even for a single-bit error. Otherwise, RAM failures may cause soft errors that result in unreliable system operation. Faulty dynamic RAM refresh circuitry is another factor to consider in diagnosing apparent RAM failures.

### RAM testing

RAMs are tested by writing a pattern into the memory, reading it back and verifying that it is unchanged. Of the many different patterns that can be used, a common one is the checkerboard. In this pattern, all the bits are set to alternating 1s and 0s. Once all memory locations have been tested, the pattern is repeated with each bit reversed, veryifying that each bit of the RAM can store a 1 and a 0. Many other patterns used to test RAMs are specifically aimed at detecting various failure mechanisms within the RAM.

No memory test can guarantee 100 per cent accuracy, even though it shows that each bit can store a 1 or a 0. RAMs can be pattern sensitive. For example, a location may well store 01010101 and 10101010 correctly but fail to store 01111110. However, even for a small RAM it would take a long time to check all locations for all possible patterns, and for a full 64 Kbyte RAM the time would be prohibitive. For this reason RAM tests are somewhat less reliable than ROM tests.

## Exercise 4.3

The program presented in Figure 4.3 implements the checkerboard test for RAM. It first writes the pattern 01010101 (55 hex) to all the memory locations to be tested and then checks that this pattern has been retained by the memory cell. After all the cells to be tested have been written and checked with this pattern the data is changed to 10101010 (AA hex) and the test repeated.

Figure 4.3   *Memory bit test*

```
0001 ;TITLE - RAMT1
0002 ;
0003 ;PURPOSE - RAM BIT TEST ROUTINE
0004 ;MEMORY IS TESTED IN 256 BYTE BLOCKS
0005 ;
0006 ;TESTS EACH MEMORY LOCATION IF NO
0007 ;ERROR IS FOUND THEN THE PROGRAM
0008 ;RETURNS TO THE MONITOR. IF THE RAM
0009 ;FAILS THEN THE MONITOR BREAK ROUTINE
0010 ;IS EXECUTED AND THE ADDRESS OF THE
0011 ;FAULTY LOCATION IS IN THE DE
0012 ;REGISTER PAIR AND THE FAIL PATTERN
0013 ;IS IN THE A REGISTER
0014 ;NO SUBROUTNES ARE CALLED AND ONLY
0015 ;CPU REGISTERS ARE USED.
0016 ;
0017 ;THE PROGRAM IS ASSEMBLED TO RUN
0018 ;IN THE PERIPHERIAL BOARD RAM BUT
0019 ;MAY BE BLOWN INTO EPROM
0020 ;
0021 ;ORG 4000H
0022 ;
0023 ;
0024 PAT1:EQU 55H
0025 PAT2:EQU 0AAH
0026 MONW: EQU 57H
0027 RST7V:EQU 2F0AH
0028 RAMST:EQU 3000H
0029 ;
0030 TEST:MVI A,PAT1 ;TEST PATTERN 1
0031 LXI H,RAMST ;GET POINTERTO MEMORY
0032 MVI B,0 ;B=MEMORY COUNTER
0033 TRY:MOV M,A ;WRITE TO LOCATION
0034 CMP M ;DID IT WRITE?
0035 JNZ ERROR ;NO
0036 INX H ;YES SO ON TO NEXT
0037 DCR B
0038 JNZ TRY ;UNTIL ALL 256 BYTES
0039 ;
0040 ;NOW TRY THE OTHER BITS
0041 ;
0042 LXI H,RAMST ;RESET POINTER ETC.
0043 MVI B,0 ;
0044 MVI A,PAT2
0045 TRY2:MOV M,A ;WRITE PATTERS
0046 CMP M ,DID IT
0047 JNZ ERROR ,NO
0048 INX H ;YES IT DID SO,
0049 DCR B ;CARRY ON
0050 JNZ TRY2:
0051 ;
0052 JMP MONW ;ALL OK.
0053 ERROR:XCHG ;SAVE FAULTY ADDRESS
0054 LHLD RST7V ;BREAKPOINT VECTOR
0055 PCHL ;JUMP TO ADDRESS IN HL
0056 END
```

If any cell fails to hold its information, or adjacent cells corrupt each other, then the program will halt and the address of the failure will be printed.

This program (Figure 4.3), written for HEKTOR, checks the user RAM between addresses 3000H and 3FFFH. Assemble, load and run this program.

Note that it is necessary to have the peripheral box plugged in as the program resides in the RAM in the peripheral chip. The program will erase anything that is in the user RAM between addresses 3000H and 3FFFH.

This program is suitable for saving in an EPROM.

The program in Figure 4.3 is longer than strictly necessary because no subroutines are called; this would involve the use of stack RAM, which may be faulty. Only the CPU registers are employed.

### Exercise 4.4

The program shown in Figure 4.3 checks that the data written into the RAM is present immediately after writing; in the case of dynamic RAM there may be a problem with the refresh circuitry. Rewrite the program of Figure 4.3 to check that the data written to the RAM is still present 100 ms afterwards.

Some faults in RAM will not be spotted by the previous test programs. They only test the ability of the memory to store data; they do not check if writing to one address corrupts the data stored in another. The effect is known as *dual address failure*. A method of checking this condition is to write the low byte of the address of a memory location to the memory as data and, having previously cleared all the other locations, to check that only one location holds that particular data.

### Exercise 4.5

Write a program to write the low byte of an address to the corresponding memory location (hint: use HL as a pointer and work in 256 byte blocks).

Then add a routine to check that the correct data appears in each location.

### ROM testing

ROMs can also fail. Such failures are more frequent when non-mask programmable types are used. A single bad bit could crash the system. Even worse, 99 per cent of it could work and 1 per cent could produce erroneous results. ROMs can be effectively tested during power-up self-test, if such tests are designed in. However, ROMs can also be tested by other techniques.

The most common technique for testing ROMs uses a check-sum. When the ROM is programmed, all of its words are added together, ignoring any carries that result. This number is *negated* (two's complement) and stored in the last (or sometimes the first) word of the ROM, so that when all words are added together (including the check-sum stored in the last or first byte) the result is zero. If the total is not zero at the end of the test sequence, then something is wrong with the ROM.

Unfortunately, the check-sum is not totally reliable. It detects any single-bit errors; however, there are many combinations of two or more errors that still produce the correct check-sum. Nevertheless, a ROM that passes a check-sum test is probably good. If the test fails, something is definitely wrong.

The HEKTOR ROMs have the check-sum loaded in the last location of each device (addresses 0FFF, 1FFF etc.) and so can be used here.

### Exercise 4.6

The program presented in Figure 4.4 checks ROM 1 for the correct check-sum. The total for the ROM less the last location is 6E (01101110) and the last location holds 92 (10010010), giving a total of 100H; as the carry is discarded the 8-bit number is 00.

Figure 4.4   *A ROM test program*

```
0001 ;TITLE —CHECKSUM
0002 ;
0003 ;PROM CHECKSUM PROGRAM. ADDS THE TOTAL BYTES
0004 ;IN PROM 1 ON HEKTOR. A TOTAL OF OTHER THAN
0005 ;ZERO DENOTES A PROM FAULT
0006 ;
0007 ROMST:EQU 0
0008 MONW:EQU 57H
0009 PRMES:EQU 30AH
0010 ;
0011  ORG 4000H
0012 ;
0013 START:LXI B,1000H ;= 4K BYTES
0014  LXI H,ROMST ;START ADDRESS OF ROM TO BE TESTED
0015  XRA A ;CLEAR A
0016 LOOP1:ADD M ;ADD BYTE
0017  INX H ;POINT TO NEXT
0018  DCR C ;DECREMET COUNTER
0019  JNZ LOOP1
0020  DCR B ;OUTER LOOP
0021  JNZ LOOP1 ;UNTIL ALL COUNTED
0022  CPI 0 ;CHECK FOR ZERO
0023  JZ ~K ;IF ZERO THEN ROM OK
0024  LXI H,MSG ;PRINT ERROR MESSAGE
0025  CALL PRMES ;
0026 OK: JMP MONW ;RETURN
0027 MSG:DB 'PROM FAIL'
0028  DB 0
0029  END
```

You can check the contents of ROM using the monitor M command.

Test this program in HEKTOR then rearrange it so that other ROMs can be checked.

### *Exercise 4.7*

Combine the PROM and RAM test programs into a single test program, get a check-sum and 'blow' the entire program into an EPROM. Test the EPROM in HEKTOR (what address should it execute at?).

## 4.5 Using special-purpose test equipment

### *Signature analyser*

Signature analysis (SA) is a technique for identifying faulty sequences in sequential logic circuits. The point under investigation is called the 'circuit node'. The signature analyser converts the long and complex data streams at the circuit node into a four-digit 'signature'. This is then compared with the signature of a known working system at the same node. When an incorrect signature is found, the signal path can be traced back until a good signature is found, thus localizing the fault.

Signatures at circuit nodes are used in a similar way to the voltage and waveform information found in analogue circuits. The node signatures are generated by a test stimulus program which exercises specific parts of the circuit in a known and repeatable way.

The essential component of a signature analysis is a pseudorandom binary sequence generator (PRBS generator). This is illustrated in Figure 4.5. The generator uses a series of D-type bistables, the D input of the following unit being fed from the Q output of the previous one.

The first stage input is fed from an exclusive–OR gate which has as its inputs the data from the node under test and the outputs from

Figure 4.5 *Pseudorandom binary sequence generator*

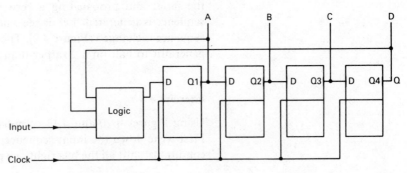

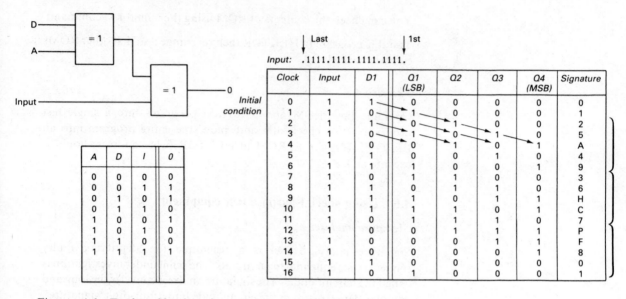

Figure 4.6  *Truth table for the exclusive–OR gate*

Figure 4.7  *PRBS example 1*

| Binary | Hex | Display |
|--------|-----|---------|
| 1010 | A | A |
| 1011 | B | C |
| 1100 | C | F |
| 1101 | D | H |
| 1110 | E | P |
| 1111 | F | U |

Figure 4.8  *Funny hex code*

stages one and four of the shift register. The truth table for the exclusive-OR (XOR) is shown in Figure 4.6. Let us examine the sequence of events illustrated by Figure 4.7, which shows the input data stream to be a series of logic ones.

*Note*: most signature analysers use a slightly different form of hexadecimal coding to that already met. This 'funny hex' or 'HP hex' is illustrated in Figure 4.8.

The initial conditions are shown at clock cycle 0. All outputs are reset to 0 so the only input to the XOR gate is the input. Therefore, D1 is also a logic 1.

After the first clock pulse the input has propagated to Q1, so the input to the XOR is now two 1s, and D1 becomes 0. On clock pulse 2 the sequence shifts right one position, allowing a further one through the XOR gate, and so on. At clock 4 Q4 has changed to a 1, so inhibiting the input and propagating a zero to the input. Notice how the sequence is generated. Let us see what happens when one bit of the sequence is changed (Figure 4.9). The zero in the sequence causes the generator to halt on a 2 rather than a 1.

### Exercise 4.8

Using Figures 4.10 and 4.11 complete the tables for the sequence. First write down the input sequence. Then, starting with the initial conditions, shift all the bits right one place before finally filling in the

Figure 4.9  *PRBS example 2*

Input:  .1111.1101.1111.1111.

| Clock | Input | D1 | Q1 (LSB) | Q2 | Q3 | Q4 (MSB) | Signature |
|---|---|---|---|---|---|---|---|
| 0 | 1 | 1 | 0 | 0 | 0 | 0 | 0 |
| 1 | 1 | 0 | 1 | 0 | 0 | 0 | 1 |
| 2 | 1 | 1 | 0 | 1 | 0 | 0 | 2 |
| 3 | 1 | 0 | 1 | 0 | 1 | 0 | 5 |
| 4 | 1 | 0 | 0 | 1 | 0 | 1 | A |
| 5 | 1 | 1 | 0 | 0 | 1 | 0 | 4 |
| 6 | 1 | 1 | 1 | 0 | 0 | 1 | 9 |
| 7 | 1 | 0 | 1 | 1 | 0 | 0 | 3 |
| 8 | 1 | 1 | 0 | 1 | 1 | 0 | 6 |
| 9 | 1 | 1 | 1 | 0 | 1 | 1 | H |
| 10 | 0 | 0 | 1 | 1 | 1 | 1 | U |
| 11 | 1 | 0 | 0 | 1 | 1 | 1 | P |
| 12 | 1 | 0 | 0 | 0 | 1 | 1 | F |
| 13 | 1 | 0 | 0 | 0 | 0 | 1 | 8 |
| 14 | 1 | 1 | 0 | 0 | 0 | 0 | 0 |
| 15 | 1 | 0 | 1 | 0 | 0 | 0 | 1 |
| 16 | 1 | 1 | 0 | 1 | 0 | 0 | 2 |

Initial condition (Clock 0)

Figure 4.10  *Exercise sheet 1*

Input:  .0000.0101.1001.0000.

| Clock | Input | D1 | Q1 (LSB) | Q2 | Q3 | Q4 (MSB) | Signature |
|---|---|---|---|---|---|---|---|
| 0 | | | | | | | |
| 1 | | | | | | | |
| 2 | | | | | | | |
| 3 | | | | | | | |
| 4 | | | | | | | |
| 5 | | | | | | | |
| 6 | | | | | | | |
| 7 | | | | | | | |
| 8 | | | | | | | |
| 9 | | | | | | | |
| 10 | | | | | | | |
| 11 | | | | | | | |
| 12 | | | | | | | |
| 13 | | | | | | | |
| 14 | | | | | | | |
| 15 | | | | | | | |
| 16 | | | | | | | |

Initial condition (Clock 0)

Figure 4.11  *Exercise sheet 2*

Input:  .0000.0000.1001.0000.

| Clock | Input | D1 | Q1 (LSB) | Q2 | Q3 | Q4 (MSB) | Signature |
|---|---|---|---|---|---|---|---|
| 0 | | | | | | | |
| 1 | | | | | | | |
| 2 | | | | | | | |
| 3 | | | | | | | |
| 4 | | | | | | | |
| 5 | | | | | | | |
| 6 | | | | | | | |
| 7 | | | | | | | |
| 8 | | | | | | | |
| 9 | | | | | | | |
| 10 | | | | | | | |
| 11 | | | | | | | |
| 12 | | | | | | | |
| 13 | | | | | | | |
| 14 | | | | | | | |
| 15 | | | | | | | |
| 16 | | | | | | | |

Initial condition (Clock 0)

Figure 4.12   *Example for exercises*
*(a) stage 1 (b) stage 2 (c) stage 3*
*(d) stage 4*

(a)
Input:  .0110.0000.0110.0101.

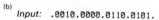

| Clock | Input | D1 | Q1 (LSB) | Q2 | Q3 | Q4 (MSB) | Signature |
|---|---|---|---|---|---|---|---|
| Initial condition   0 | 1 | 1 | 0 | 0 | 0 | 0 | 0 |
| 1 | 0 | | 1 | 0 | 0 | 0 | 1 |
| 2 | 1 | | | | | | |
| 3 | 0 | | | | | | |
| 4 | 0 | | | | | | |
| 5 | 1 | | | | | | |
| 6 | 1 | | | | | | |

(b)
Input:  .0010.0000.0110.0101.

| Clock | Input | D1 | Q1 (LSB) | Q2 | Q3 | Q4 (MSB) | Signature |
|---|---|---|---|---|---|---|---|
| Initial condition   0 | 1 | 1 | 0 | 0 | 0 | 0 | 0 |
| 1 | 0 | 0 | 1 | 0 | 0 | 0 | lost  1 |
| 2 | 1 | | 0 | 1 | 0 | 0 | 2 |
| 3 | 0 | | | | | | |
| 4 | 0 | | | | | | |
| 5 | 1 | | | | | | |
| 6 | 1 | | | | | | |

(c)
Input:  .0110.0000.0110.0101.

| Clock | Input | D1 | Q1 (LSB) | Q2 | Q3 | Q4 (MSB) | Signature |
|---|---|---|---|---|---|---|---|
| Initial condition   0 | 1 | 1 | 0 | 0 | 0 | 0 | 0 |
| 1 | 0 | 0 | 1 | 0 | 0 | 0 | lost  1 |
| 2 | 1 | 1 | 0 | 1 | 0 | 0 | 2 |
| 3 | 0 | | | | | | |
| 4 | 0 | | | | | | |
| 5 | 1 | | | | | | |
| 6 | 1 | | | | | | |

(d)
Input:  .0110.0000.0110.0101.

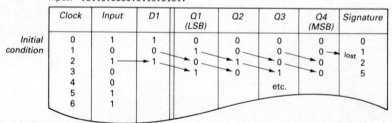

| Clock | Input | D1 | Q1 (LSB) | Q2 | Q3 | Q4 (MSB) | Signature |
|---|---|---|---|---|---|---|---|
| Initial condition   0 | 1 | 1 | 0 | 0 | 0 | 0 | 0 |
| 1 | 0 | 0 | 1 | 0 | 0 | 0 | lost  1 |
| 2 | 1 | 1 | 0 | 1 | 0 | 0 | 2 |
| 3 | 0 | | 1 | 0 | 1 | 0 | 5 |
| 4 | 0 | | | | etc. | | |
| 5 | 1 | | | | | | |
| 6 | 1 | | | | | | |

D1 column. Figure 4.12 (a–d) shows the process stage by stage. Remember if Q1 and Q4 are both at 0 or 1 then the input signal is passed to the first stage. When they are different input is inhibited.

A complete signature analyser requires three input signals and will consist of a 16-bit shift register and associated logic.

The shift register and XOR inputs are shown in Figure 4.13.

Figure 4.14 shows the sequence of events which occurs on signature acquisition. Clock, start and stop signals are specified for a 0 to 1 transition.

Figure 4.13 *Input arrangements for a commercial SA*

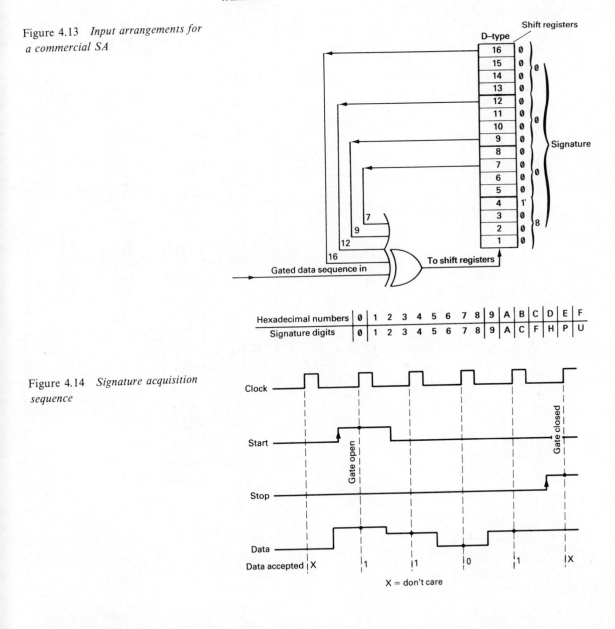

Figure 4.14 *Signature acquisition sequence*

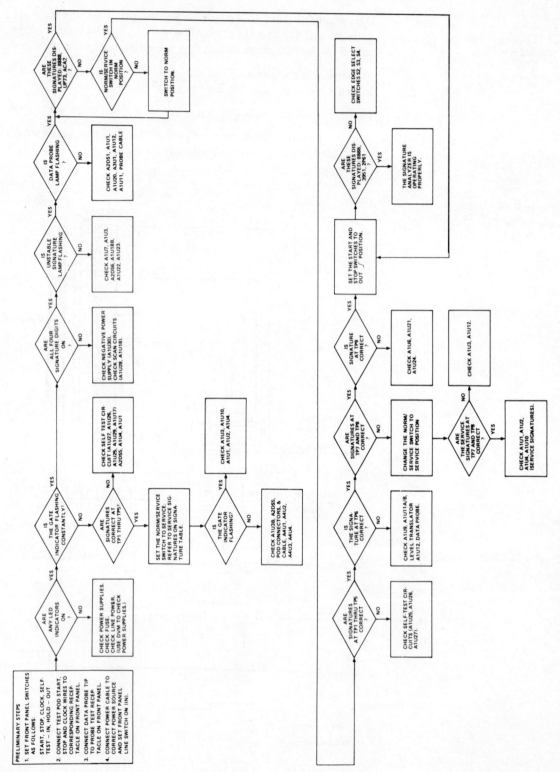

Figure 4.15   *A typical SA fault-finding guide* (*source:* Hewlett-Packard 5004 Manual)

| PINS | U1 | U2 | U3 | U4 | U5 | U6 | U7 | U8 | L 9 | U10 | PIN |
|---|---|---|---|---|---|---|---|---|---|---|---|
| 1 N | 472A | 5A22 | | 1H08 | 5A22 | 472A | F517 | UCP9 | 472A | 7CA7 | 1 |
| S | 472A | 94A3 | | H389 | 94A3 | 472A | P7AA | PF43 | 472A | 7CA7 | |
| 2 N | A326 | A326 | | 09P3 | 472A | 472A | 0000 | P36F | 3F3H | 7CA7 | 2 |
| S | A326 | A326 | | 09P3 | 472A | 472A | 0000 | P36F | 3F3H | 7CA7 | |
| 3 N | P40F | UCP9 | | 1H08 | | 472A | 823H | CFF3 | 7CA7 | 472A | 3 |
| S | P40F | PF43 | | H389 | | 472A | A080 | Ac69 | 7CA7 | 472A | |
| 4 N | 464F | UCP9 | | UCP9 | P40F | 472A | 4C4F | CFF3 | 472A | 0000 | 4 |
| S | 464F | PF43 | | PF43 | P40F | 472A | 125P | AC69 | 472A | 0000 | |
| 5 N | 13F7 | UCP9 | | UCP9 | 5829 | 596F | 0F66 | 66P0 | 596F | 472A | 5 |
| S | 13F7 | PF43 | | PF43 | A427 | 596F | 5574 | 6606 | 596F | 472A | |
| 6 N | 4PF9 | 3P06 | | UCP9 | H4U0 | 0000 | 0147 | UCP9 | 1P46 | 0000 | 6 |
| S | 4PF9 | 62CF | | PF43 | 6H73 | 42U6 | 0000 | PF43 | 1P46 | 0000 | |
| 7 N | 09P3 | 0000 | | 0000 | 0000 | 0000 | 0000 | 0000 | | 0000 | 7 |
| S | 09P3 | 0000 | | 0000 | 0000 | 0000 | 0000 | 0000 | | 0000 | |
| 8 N | 0000 | 0000 | | C445 | 66P0 | 0000 | H4U0 | 472A | 0000 | 13F7 | 8 |
| S | 0000 | 0000 | | 1669 | 6606 | 0000 | 6H73 | 472A | 0000 | 13F7 | |
| 9 N | 0000 | 5829 | | 5829 | 5829 | 0000 | HAU1 | FUFU | | 54PH | 9 |
| S | 0000 | A427 | | A427 | A427 | | HAU1 | FUFU | | 54PH | |
| 10 N | | 4PF9 | | P40F | P40F | F944 | 0F66 | 0863 | | 464F | 10 |
| S | | 4PF9 | | P405 | P40F | CFU5 | 5574 | 0863 | | 464F | |
| 11 N | 5829 | 4PF9 | | 5829 | P40F | AUF8 | 4596 | 7CA7 | | 0166 | 11 |
| S | A427 | 4PF9 | | | A427 | HHH5 | 4596 | 7CA7 | | 0166 | |
| 12 N | 3P06 | 4PF9 | 54PH | 1H08 | 5A22 | 2CAU | 2946 | 7A33 | | 0166 | 12 |
| S | F61C | 4PF9 | 54PH | H389 | 94A3 | 6PAH | 2946 | 7A33 | | 0166 | |
| 13 N | C445 | A326 | 0166 | 1H08 | P36F | 1501 | 90FP | 4596 | | A446 | 13 |
| S | 2946 | A326 | 0166 | H389 | P36F | 1417 | 90FP | 4596 | | A446 | |
| 14 N | 1H08 | 472A | | 472A | 472A | 472A | 472A | 472A | | 472A | 14 |
| S | H389 | 472A | | 472A | 472A | 472A | 472A | 472A | | 472A | |
| 15 N | 5A22 | | | | | | | | | 472A | 15 |
| S | 94A3 | | | | | | | | | 472A | |
| 16 N | 472A | | | | | | | | | 472A | 16 |
| S | 472A | | | | | | | | | 472A | |

| PIN | U11 | U12 | U13 | U14 | U15 | U16 | U17 | U18 | U19 | U20 | PIN |
|---|---|---|---|---|---|---|---|---|---|---|---|
| 1 N | 7CAF | | | | | | 90FP | 0000 | 6892 | | 1 |
| S | 7CAF | | | | | | 90FP | 0000 | 802C | | |
| 2 N | 7CAF | | | | | | HH53 | | 443F | | 2 |
| S | 7CAF | | | | | | HH53 | | 80CH | | |
| 3 N | 3F8H | | 75U6 | 75U6 | 75U6 | 75U6 | 75U6 | | 2CHF | | 3 |
| S | 3F8H | | 0261 | 0261 | 0261 | 0261 | 0261 | | 99U2 | | |
| 4 N | 3F8H | 0000 | A096 | A096 | A096 | A096 | | 4C78 | 27U3 | | 4 |
| S | 3F8H | 0000 | 92PC | 92PC | 92PC | 92PC | | 4C78 | 9H02 | | |
| 5 N | 3F8H | 472A | 3A0U | 3A0U | 3A0U | 3A0U | 0863 | | 25CF | 069C | 5 |
| S | 3F8H | 472A | 9664 | 9664 | 9664 | 9664 | | 25CF | 0HAH | | |
| 6 N | 7CA7 | | FU22 | FU22 | FU22 | FU22 | A096 | 7661 | 78CP | | 6 |
| S | 7CA7 | | C152 | C152 | C152 | C152 | 92PC | 7661 | PH0C | | |
| 7 N | 0000 | | | | | | 0000 | 5U8U | P73H | | 7 |
| S | 0000 | | | | | | 0000 | 5U8U | CH2U | | |
| 8 N | U36U | | 0000 | 0000 | 0000 | 0000 | FU22 | 0000 | | | 8 |
| S | 6P6F | | 0000 | 0000 | 0000 | 0000 | C152 | 0000 | | | |
| 9 N | C445 | | 0000 | 0000 | 0000 | 0000 | 7A33 | 472A | | 9 | |
| S | 2946 | | 0000 | 0000 | 0000 | 0000 | 7A33 | 472A | | | |
| 10 N | C445 | | 0000 | 0000 | 0000 | 0000 | | | | | 10 |
| S | 2946 | | 0000 | 0000 | 0000 | 0000 | | | | | |
| 11 N | 472A | | FH33 | C826 | F94H | AUF8 | 3A0U | 0000 | | | 11 |
| S | 472A | | FUAU | PU7H | CFU5 | HHH5 | 9664 | 0000 | | | |
| 12 N | 3F8H | 3F8H | 1501 | 6C7H | 929A | 47SF | 29PP | 472A | | | 12 |
| S | 3F8H | 3F8H | 1417 | 5553 | U242 | 3003 | 29PP | 472A | | | |
| 13 N | 7CAF | 7CA7 | APH9 | 5F97 | 2535 | 9FU2 | | 472A | | | 13 |
| S | 7CAF | 7CA7 | 3AAA | C822 | U600 | 0000 | | 472A | | | |
| 14 N | 472A | | 54F8 | 94FH | 52A7 | 2CAU | 472A | 0000 | 0000 | | 14 |
| S | 472A | | UPUF | 7CCH | 67A8 | 6PAH | 472A | 0000 | 0000 | | |
| 15 N | | | 0000 | 0000 | 0000 | 0000 | | 0000 | 0000 | | 15 |
| S | | | 0000 | 0000 | 0000 | 0000 | | 0000 | 0000 | | |
| 16 N | | | | | | | | 472A | 472A | | 16 |
| S | | | | | | | | 472A | 472A | | |

N = NORMAL
S = SERVICE position of S7.

To get the signatures given in this table, set the two 5004A's controls as follows:

5004A Being Tested
LINE:OFF; START:OUT; STOP:OUT; HOLD:OUT; SELF-TEST:IN.

5004A Used to Test
Same as above except SELF-TEST:OUT

Make the connections shown between the two 5004A's.

**5004A USED FOR TESTING**

| PIN | U21 | U22 | U23 | U24 | U25 | U26 | U27 | U28 | U29 | U30 | PIN |
|---|---|---|---|---|---|---|---|---|---|---|---|
| 1 N | 0147 | | | | F61C | 0000 | HH53 | | 54PH | | 1 |
| S | 596F | | | | F61C | 0000 | HH53 | | 54PH | | |
| 2 N | 0147 | | | | 0000 | 0000 | 0000 | | 0166 | | 2 |
| S | 596F | | | | 0000 | 0000 | 0000 | | 0166 | | |
| 3 N | 94FH | | | 2CAU | 0000 | 0000 | 0000 | | A446 | | 3 |
| S | 7CCH | | | 6PAH | 0000 | 0000 | 0000 | | A446 | | |
| 4 N | 5F97 | 29PP | | 9FU2 | 2946 | | | | HAU1 | | 4 |
| S | C822 | 29PP | | 7282 | 2946 | | | | HAU1 | | |
| 5 N | 6C7H | 7A33 | | 47F5 | | | | | | | 5 |
| S | 5553 | 7A33 | | 3003 | | | | | | | |
| 6 N | C826 | 14HA | | AUF8 | | | | | | | 6 |
| S | PU7H | 7782 | | HHH5 | | | | | | | |
| 7 N | 0000 | 29H7 | | 0000 | | | | | | | 7 |
| S | 0000 | P5U1 | | 0000 | | | | | | | |
| 8 N | | | | | 4596 | 29PP | 3A9A | | | | 8 |
| S | | | | | 4596 | 29PP | 3A9A | | | | |
| 9 N | | 207P | | | | | 7A33 | H10F | | | 9 |
| S | | A5C9 | | | | | 7A33 | H10F | | | |
| 10 N | 54F8 | F2P7 | F2P7 | 52A7 | 2946 | 7A33 | H10F | | 29PP | | 10 |
| S | UPUF | OFC1 | OFC1 | 67A8 | | | | | 29PP | | |
| 11 N | APH9 | 0000 | | 2535 | FUFU | 0863 | 0108 | | 0863 | | 11 |
| S | 3AAA | 0000 | | U600 | FUFU | 0863 | 0108 | | 0863 | | |
| 12 N | 1501 | 472A | 207P | 929A | F61C | 0000 | HH53 | | | | 12 |
| S | 1417 | 472A | A5C9 | U242 | F61C | 0000 | | | | | |
| 13 N | FH33 | 29PP | 29H7 | F94H | | | | | | | 13 |
| S | FUHU | 29PP | P5U1 | CFU5 | | | | | | | |
| 14 N | 472A | | | 472A | 0108 | 0000 | 0863 | | | | 14 |
| S | 472A | | | 472A | 0108 | 0000 | 0863 | | | | |
| 15 N | | | 14HA | | | | | | | | 15 |
| S | | | 7782 | | | | | | | | |
| 16 N | | | | | | | | | | | 16 |
| S | | | | | | | | | | | |

Figure 4.16   *Connecting HEKTOR to an SA (source:* Hewlett-Packard 5004 Manual)

1 Start signal goes high.
2 On the first clock pulse after this transition, the state of the circuit node is stored; then on each clock pulse the data is shifted and new data stored as explained above, until
3 The stop signal is received, when the contents of the shift register are displayed as a four-digit 'funny hex' number.

As stated previously, signatures are used in digital circuits as voltages are in analogue circuits.

Figure 4.15 shows a typical fault-finding guide based on signatures, and Figure 4.16 shows an actual set-up for fault-finding on the Hewlett-Packard 5004A signature analyser.

To implement the signature analysis to a high level it has to be 'designed in' to the equipment. However, if a known good system is

Figure 4.17  *Some HEKTOR signatures*

| PIN # | IC # 13 | 7 | (2) | 12 |
|---|---|---|---|---|
| 1. | IUFP | P825 | 0000 | 0000 |
| 2. | 0000 | HCHC | IUFP | 3C1U |
| 3. | 0000 | 32AA | 4AC4 | 3C1U |
| 4. | 0000 | 01C0 | * | 93C2 |
| 5. | IUFP | 3900 | * | 93C2 |
| 6. | IUFP | P776 | 0000 | 6C48 |
| 7. | 0000 | 4AC4 | 0000 | 6C48 |
| 8. | 0000 | 0000 | | 68PC |
| 9. | IUFP | 2123 | | 68PC |
| 10. | 0000 | FC32 | | 0000 |
| 11. | IUFP | 13U3 | | IUFP |
| 12. | 3C1U | A0UP | | 6445 |
| 13. | 93C2 | U0P6 | | 6445 |
| 14. | 6C48 | AA80 | | F4HA |
| 15. | 68PC | 1029 | | F4HA |
| 16. | 6445 | IUPF | | C851 |
| 17. | F4HA | | | C851 |
| 18. | C851 | | | 7040 |
| 19. | 7040 | | | 7040 |
| 20. | 0000 | | | IUFP |
| 21. | 49AA | | | |
| 22. | 73P4 | | | |
| 23. | FC32 | | | |
| 24. | 13U3 | | | |
| 25. | A0UP | | | |
| 26. | U0P6 | | | |
| 27. | AA80 | | | |
| 28. | HHH5 | | | |
| 29. | 1945 | | | |
| 30. | IUFP | | | |
| 31. | IUFP | | | |
| 32. | IUFP | | | |
| 33. | 0UP7 | | | |
| 34. | * | | | |
| 35. | IUFP | | | |
| 36. | IUFP | | | |
| 37. | 0000 | | | |
| 38. | 0000 | | | |
| 39. | 0000 | | | |
| 40. | IUFP | | | |

* = UNSTABLE SIGNATURE

available then a test-and-compare method can be used, and some signature analyser units have the facility for storing a signature obtained from one unit under test for later comparison with that from another unit.

Figure 4.17 is a list of some signatures obtained from HEKTOR, using the program of Figure 4.18.

Figure 4.18 *A program to exercise HEKTOR for SA*

```
TTEKTRONIX  8080/8085 ASM V3.3  SIG. ANL

00002            ;            PROGRAM :-SIGNATURE ANALYSIS TEST PROGRAM
00003            ;            VERSION :-1.0
00004            ;
00005            ;
00006            ;
00007            ;            PROGRAMMER :-P. D. S.
00008            ;
00009            ;            DATE :- MAY 1982
00010            ;
00011            ;
00012            ;            INPUT ;-NOTHING
00013            ;
00014            ;
00015            ;            OUTPUT ;-NOTHING
00016            ;
00017            ;
00018            ;            REGISTERS USED :-A,H,L
00019            ;
00020            ;
00021            ;
00022                              PROGRAM DESCRIPTION
00023            ;                 ===================
00024            ;            THIS PROGRAM EXERCISES THE ADDRESS BUS OF
00025            ;            OF THE HEKTOR COMPUTER. DUE TO THE PROBLEMS
00026            ;            OF MULTIPLEXING THE DATA AND ADDRESS
00027            ;            BUS IT IS NECESSARY TO PRODUCE A STOP/START
00028            ;            SIGNAL FROM THE IO/M LINE.
00029            ;
00030            ;
00031            ;
00032 0000  210000   START   LXI    H,0      ;SET START ADDRESS FOR TEST
00033 0003  7E       LOOP    MOV    A,M      ;EXERCISE ADDRESS BUS
00034 0004  23               INX    H        ;NEXT ADDRESS
00035 0005  7D               MOV    A,L      ;TEST FOR ONE
00036 0006  B4               ORA    H        ;CIRCUIT OF THE ADDRESSES
00037 0007  C20300  >        JNZ    LOOP     ;NOT YES
00038 000A  D300             OUT    0        ;START/STOP SIGNAL
00039 000C  C30300  >        JMP    LOOP     ;AND CONTINUE FOR EVER
00040                        END

EKTRONIX  8080/8085 ASM V3.3  SYMBOL TABLE

Scalars

A------0007   B------0000   C------0001   D------0002
E------0003
H------0004   L------0005   M------0006   PSW----0006
SP-----0006

%(default) Section (000F)

LOOP---0003   START -- 0000
```

*Connecting the signature analyser*

The signature analyser (SA) has to be provided with three timing signals, i.e. start, stop and clock. The simplest way to ensure a repetitive sequence is to open-circuit the data bus and 'force' a NOP instruction on it. This allows the address bus to cycle from 0000 to FFFF passing all addresses on its way, so all address lines and related logic are exercised. Owing to the multiplexing of the address and data on the 8085 bus, this is not so easy. The program shown in Figure 4.18 exercises each address line and also provides start and stop signals.

Now that a sequence is established the start and stop inputs to the signature analyser can be provided by IO/$\overline{\text{M}}$ and by selecting similar edges for each signal.

The choice of signal for the clock is rather limited. As the data and address buses are multiplexed the normal clock is not usable, but the $\overline{\text{RD}}$ signal appears for every instruction and this may be used for the signature analyser clock.

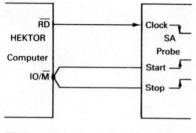

Figure 4.19   *Connecting HEKTOR to a signature analyser*

### Exercise 4.9

Connect up the signature analyser as shown in Figure 4.19.

### Exercise 4.10

Simulate a fault condition by, say, short-circuiting two address lines, and see how this affects the signatures obtained.

If two circuit nodes produce the same signature then they are probably connected together. It is then up to the technician to decide if they *should* be connected together.

### Logic analysers

A logic analyser is a test instrument capable of capturing binary signals on from 8 to 48 inputs simultaneously. The inputs are sampled a number of times (typically 256 times) in quick succession under the control of a clock signal. The resultant input data is stored in the *data acquisition* memory. The data acquisition sequence is initiated by a trigger condition which can be provided by external means, or by the recognition of a particular binary pattern on some or all of the inputs. Data patterns are recorded both before and after the trigger condition. A microprocessor system is then programmed to assess and display the stored data on a cathode ray tube.

Figure 4.20 shows the block diagram of a typical logic analyser.

Input digital signals are applied continuously via the acquisition probes or pods to the parallel data input circuit (block A). Samples are taken of the incoming data under the control of the clock gates

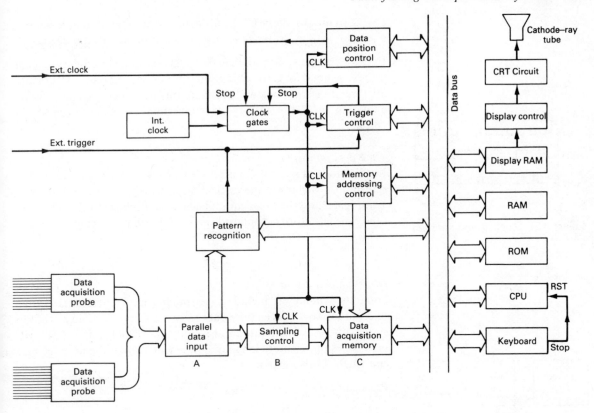

Figure 4.20  *Block diagram of a logic analyser*

(block B) and the sampled data is stored in the data acquisition memory on a first-in, first-out (FIFO) basis. The trigger control block ends the data acquisition phase upon receipt of a trigger signal from the pattern recognition circuits or from an external source.

Once the data has been stored (block C) then the CPU will control the transfer of data to the display RAM using the data position and memory addressing control circuits.

Several modes of display are possible. Those in common use are outlined below.

*Binary number display*
The stored digital signals are interpreted as 0 or 1 and are displayed as such using a character generator. There is usually an ability to select either a positive or a negative logic convention.

| Successive | 00101100 | 11011100 | 01011010 | ← CRT with |
| samples ↓ | 00101100 | 11011100 | 01011011 | 24-bit |
| | 11111111 | 11011100 | 01011100 | display |

Owing to the size of the CRT only a limited number of samples can be shown at a time, so facilities are provided to 'scroll' the display.

Facilities may also be provided to display the binary numbers in octal or hexadecimal form for ease of interpretation; on some instruments all three presentations are available on the screen together. Some analysers have the facility of converting the binary to ASCII characters for display.

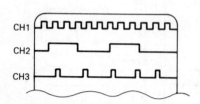

Figure 4.21   *Logic analyser displays: time domain*

### Time domain display

In this mode (Figure 4.21) a number of traces are provided on the screen. For each trace, the vertical displacement corresponds to logic 1 or logic 0 and the horizontal displacement relates to the sampling time. The waveforms displayed show *all transitions with ideal vertical edges.*

### Disassembled display

Some logic analysers include a personality module for a particular microprocessor. This enables data to be read from the data bus of the system under test and, after disassembly, the display on the CRT will be a series of assembly language statements. Often the hexadecimal equivalent is also given.

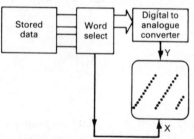

Figure 4.22   *Logic analyser displays: graph display*

For example:

```
3E02       LD    A, 02
47         LD    B, A
218040     LD    HL, 4080
```

### Graph display

This mode of display (Figure 4.22) involves converting a group of stored binary signals to an analogue signal using a DAC. Successive samples of the chosen group of binary signals are then displayed to form a graph.

Graphical displays can be useful to highlight data flow sequences.

### Map display

The map mode of display (Figure 4.23) is achieved by providing analogue inputs to both the X and Y deflection systems. If the sequence of events under observation is repeated, then the map display gives a very rapid overview of the activity.

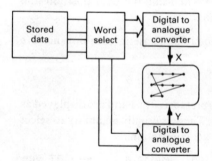

Figure 4.23   *Logic analyser displays: map display*

### Specifications

Logic analysers can be classified as:

1   Logic state analysers (synchronous sampling), or
2   Timing analysers (asynchronous sampling).

The application for both types can be:

1 General digital systems
2 Microprocessor systems generally, or
3 Specific microprocessor systems.

Many logic analysers provide for both logic state and timing analysis.

The capacity of a logic analyser is measured in terms of:

1 Data acquisition rate
2 Storage ability
3 Number of input channels
4 Display capability
5 Trigger modes
6 Glitch capture capability

Table 4.1 compares the performance of several different types of logic analyser.

*Sampling the data*
Logic analysers sample incoming data at discrete times. Binary states are stored in memory at the edge of a clock signal.

Figure 4.24 shows the outline idea. Ch0 to ChN sample the data in parallel, which is then stored in a sequential access memory. The 'threshold set' control allows the logic 0 and 1 levels to be set to suit the equipment.

The process is continuous, with old data being overwritten by new until a 'trigger' is received, when the data which is coming in is stored in the memory and returned.

Figure 4.24 *Data sampling*

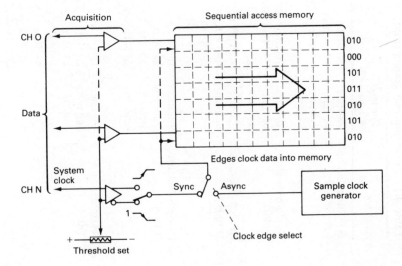

**Table 4.1**  *Logic analyser comparisons*

| Analyser | Channels & type | Clocks | Trigger modes | Memory (CH×bits) | Display capability | Glitch capture | Reference memory | Trigger delays |
|---|---|---|---|---|---|---|---|---|
| E-H International LA 1850 | 18-state or time domain | 50 MHz int. or ext. | Pre- and post-trigger | 18×512 | 18-CH timing 16-CH bin., oct. or hex. map display | 5 ns | 16×51 | By no. of clocks or no. of trigger events |
| Gould-biomation K100D | 16-state or time domain | 100MHz int. 70 MHz ext. | Pre- and post-trigger | 16×1024 | 16-CH timing 16-CH bin., oct. or hex. | 5 ns | 16×1024 | By no. of clocks or no. of triggers |
| Hewlett-Packard 1610 B | 32-state domain | Accepts up to 3 ext. clocks each ≯ 10 MHz | Pre- and post-trigger | 32×64 | Bin., oct. or hex. | None | 32×64 | By no. of clocks or no. or triggers |
| Hewlett-Packard 1615 A | 8 time or state plus 16-state domain | Int. 2 Hz – 20 MHz ext. DC – 20 MHz | Pre- and post-trigger comprehensive range of trigger levels | 24×256 8×256 for glitches | 8-CH timing 16-CH bin., oct. or hex. or 24 bin. oct. or hex. | 5 ns | None | By no. of clocks or no. of triggers |
| Tektronix 7D01 | 16-time and state domain | 16-CH at 20 MHz 4-CH at 100 MHz 8-CH at 50 MHz Int. 100 MHz Ext. 50 MHz | Pre-, centre or post trigger | 8×252 | Hex., bin. or oct. ASCII, GP18 mnemonics 16-CH timing | 15 ns 5 ns by option | 8×252 | Delay by clock pulse |
| Tektronix 7D02 | 8-time 28 state | 10 MHz 50 MHz option | Pre-, post- & glitch trigger 4 levels of triggering | 28×256 | 24 line × 32 characs. 28-CH data bin., hex., oct. or ASCII disassembly | 5 ns | 28×256 | By no. of clocks or no. of triggers |
| Tektronix 308 | 8-time & state domain | 20 MHz Inst. or ext. | Pre- or post-trigger | 8×252 | 8-CH timing hex., bin. & oct. or ASCII | 5 ns | 8×252 | By clock pulse or trigger event |

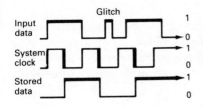

Figure 4.25 *Synchronous sampling*

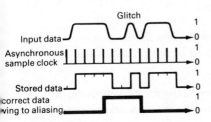

Figure 4.26 *Asynchronous sampling*

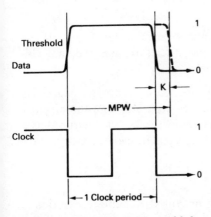

Figure 4.27 *Minimum pulse width for data capture*

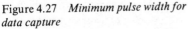

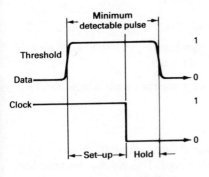

Figure 4.28 *Data set-up delay*

## Synchronous sampling

In the synchronous mode (Figure 4.25), data is sampled by a clock from the system under test. This makes a record of the states of the input lines at sequential system clock times. Narrow pulses (glitches) between clock pulses are rejected. Logic state analysers only sample data synchronously; logic timing analysers sample asynchronously.

## Clock qualifier

The clock qualifier allows data to be sampled selectively for storage in memory.

## Asychronous sampling

Asynchronous sampling (Figure 4.26), in which mode an independent clock in the logic analyser is used, can give more resolution than synchronous sampling. Incoming data is sampled asynchronously with respect to the system clock, timing information is obtained because the logic analyser clock can sample faster than the system clock.

## Aliasing

Aliasing results when the sample clock rate is slower that the data rate.

## Glitch latch

The glitch latch detects pulses to narrow to meet minimum pulse-width criteria. Any transition that occurs between sample clocks is displayed as a one-clock-period-wide pulse during the next clock interval.

## Limitations

As with all logic circuits, the logic analyser has its speed or frequency limitations for capturing the incoming signal. Figures 4.27 and 4.28 illustrate these points.

## Synchronous data timing

The incoming data must be present for set-up and hold times with respect to the clock edge.

$t_s$ (set-up time): time data must be present before the clock transition.
$t_h$ (hold time): time data must be present after the clock transition. (0 ns is optimum).

The theoretical maximum valid clock rate at which data can be acquired and stored in memory is:

$$\frac{1}{t_s + t_h}$$

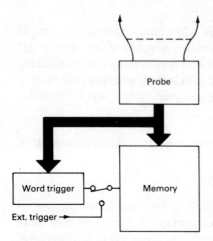

Figure 4.29   *Separate data word and trigger word*

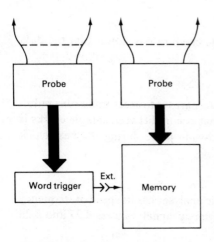

Figure 4.30   *Using the data word as a trigger.*

### Asynchronous data timing

Because the timing relationship between the logic analyser's internal clock and incoming data cannot be defined, set-up and hold time specifications have no meaning. Instead, *minimum pulse width* (MPW) is used to indicate the narrowest pulse that can be detected with 100 per cent certainty.

$$\text{MPW} = 1 \text{ clock} + k$$

where $k$ is some specified constant period.

### Trigger sources

Unique single-channel data framing events are frequently available. Combinatorial triggering (bit pattern or word recognition) is a useful extension of this idea.

A parallel word recognizer produces a trigger by comparing the logic state of each incoming data channel with an operator-selected state and combining the results for all parallel channels in an AND gate. A trigger is produced whenever a selected word occurs. The operator can select a HI, LO or X (don't care) state for each channel (see Figure 4.29).

The word recognizer most commonly sees the data input via the analyser's acquisition circuitry and, therefore, has the same bit capacity as the analyser has channels.

An alternative arrangement for the word recognizer places its acquisition and word trigger circuitry separate from the memory acquisition circuitry. The word recognition probes can then produce a word trigger from source data that is different from the memory source data (see Figure 4.30).

### Synchronous word recognition

In this mode a system clock transition is used to enable recognition of the selected data pattern. For a trigger to be generated, the selected word must be true at the edge of a system clock. Spurious pulses or 'glitches' between clocks are thus ignored.

### Asychronous word recognition

When asychoronous word recognition is selected, a trigger is produced any time the selected word occurs.

Facilities are usually provided to examine the events which occurred before, as well as after, trigger.

### Pre-trigger data

The trigger stops data acquisition and storage and the memory holds data stored before the trigger was received. This unique characteristic allows you to look backward in time to examine the logic states or

Figure 4.31   *Looking backwards in time*

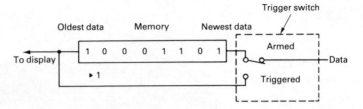

timing relationships that preceded the trigger. The longer the memory register, the farther back you can see.

In the example in Figure 4.31 data is continually stored until the trigger is received, at which time the switch is moved to the 'triggered' position and the data is held in the register.

*Trigger delay*
Because data sequences can be very long and memory capacity is finite, it is desirable to have some means of positioning the memory 'window' (Figure 4.32).

Figure 4.32   *Delayed trigger*

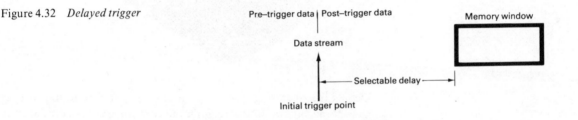

Virtually any data window position downstream from the initial trigger point can be selected with digital delay. A delay-by-events counter, started by the initial trigger, counts a preset number of events and triggers the logic analyser. The preset number can be any operator-selected value. The events are usually system clock pulses.

### Exercise 4.11   *Connecting a logic analyser to HEKTOR*

*Note*:   It is recommended that connections be made to a suitable edge connector connected to the bus expansion connector.

The main problem encountered with logic analysers is that so many probes have to be connected to the circuit. Where connections to chips have to be made a logic clip can ease the problem (Figure 4.33).

In the case of HEKTOR all connections are brought out to an edge connector, which considerably simplifies connections. Additionally, at this point in the circuit the address bus has been demultiplexed, so the full 16-bit address bus is available.

Figure 4.33   *Using a logic clip*

Details for operating individual instruments can be found in user handbooks but, in general, eight inputs are connected to the data bus and sixteen inputs are connected to the address bus.

Tho other connections are required – the common connection and the system clock.

With all the connections made, set the trigger word to a known value (31 is suggested as this instruction, LXI SR, occurs in the monitor) and then press START on the logic analyser *whilst* holding HEKTOR's RESET button pressed. Then release the RESET button; after a short pause the analyser should show each bus 'transaction'.

### Exercise 4.12

Run a known program. Set up a new trigger word, then run a trace on the analyser.

Why do some addresses appear for more than one cycle whilst others appear for only one?

Further work with the analyser will depend on its facilities. It is a powerful tool when developing a system, but somewhat expensive to use as a standard piece of test gear.

### 4.6   Conclusion

In this chapter we have found that faults can be either software or hardware induced. If a system has been in operation for some time then the fault is likely to be due to a component failure rather than incorrect assembly or a solder splash; however, these possibilities cannot be ruled out.

The testing of a microprocessor-based system presents its own specialized problems, but provided the kernel of the system is operational then a few well-chosen and written diagnostic routines in a dedicated ROM can go a long way towards isloating the faulty area.

If possible try to test the processor system alone as a processor, and when it is proved that this is working then the peripherals can be tested if the fault still persists. Additionally in this arrangement the processor can exercise the peripherals. Conventional test gear is of limited use in a microprocessor-based system, as we have noted, because of the bus structure and the fact that the system is a 'state device' rather than an analogue device. Further, some processors (such as the 8085) have multiplexed data/address/control highways, which make sorting the signals impossible with a CRO.

Special-purpose test equipment is available to solve these problems, and each type has its strengths and limitations. If designed in, signature analysis is a powerful tool and possibly the simplest to use in the field.

A data analyser will bridge the gap between SA and the full-blown logic analyser, which is best left to the development area.

Much time can be saved if the service technician has a good knowledge of the system as a whole, and the accumulation of reliability data (stock faults) is very worth while.

Taken logically, fault-finding a microprocessor-based system should present no more difficulty than working on an analogue system of the same complexity.

# Chapter 5  Case studies in microprocessor system development

*Objectives of this chapter   When you have completed this chapter you should be able to recognize the steps in the development of a microprocessor system as:*

*1   Initial specification of complete system.*
*2   Formulation of measures of cost-effectiveness.*
*3   Derivation of a system flow chart.*
*4   Definition of hardware/software implementation.*
*5   Selection of necessary electronic devices, controller, etc., and the necessary interfacing devices.*
*6   Definition of boundary constraints for the microcomputer program, e.g. timing, address space.*
*7   Development of program flow chart.*
*8   Writing of program.*
*9   Testing of program.*
*10   Modification, where necessary.*
*11   Simulation/emulation of system.*
*12   Further modification, where necessary.*
*13   Running of hardware and firmware prototype.*
*14   Iteration, as necessary.*

*Also, you should be able to recognize that the ratio of testing and debugging period to initial development time rises significantly with complexity of task.*

## 5.1   Introduction

This chapter is intended to integrate previous work and involve you in the production of a working microprocessor-based system.

In general, complete programs are not given in this section but flow charts and suggestions for writing the programs are. This is because there is no unique solution for most problems and if possible a number of alternatives should be tried.

Two projects are described here. The first is a microprocessor-operated minidrill suitable for drilling printed circuit boards, and is described in some detail. The second is a remote data logging system (weather station), which is a larger undertaking and is probably more suited to a small team of development engineers/technicians than to an individual. However, in both cases the techniques and processes are applicable to any project and other ideas may well present themselves to the student.

The development of a microprocessor-based system is influenced by a large number of factors, both economic and technological. This discussion assumes that appropriate market surveys etc. have been completed and a need for this product has been established.

In any development, a *top down* approach to the planning of the work is recommended. This means starting with a general overall description and specification of the *end product* and then refining each question and problem until the desired result is obtained.

## 5.2   Microprocessor-operated minidrill

The end product of this development is to be a programmable minidrill suitable for drilling printed circuit boards, either on a 'learn and repeat' basis or for drilling specific patterns based on a specified co-ordinate, e.g. a 40-pin DIL package with the co-ordinate as one corner pin specified. The maximum size of board to be accommodated is 100 mm × 160 mm (single Eurocard).

Note that this is a description of what the device must do; no indication is given as to how it is to be done or how it is to be implemented.

Before any decisions are made on the implementation, it is necessary to know some tolerances within which the machine is required to work. Figure 5.1 gives an extended specification for the drill.

Figure 5.1   *Specification of a programmable minidrill*

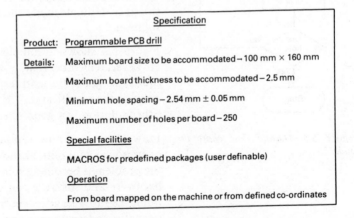

> **Specification**
>
> Product:   Programmable PCB drill
>
> Details:   Maximum board size to be accommodated – 100 mm × 160 mm
>
> Maximum board thickness to be accommodated – 2.5 mm
>
> Minimum hole spacing – 2.54 mm ± 0.05 mm
>
> Maximum number of holes per board – 250
>
> Special facilities
>
> MACROS for predefined packages (user definable)
>
> Operation
>
> From board mapped on the machine or from defined co-ordinates

From this specification the system flow chart can be produced (Figure 5.2). This shows the three possible operating modes for the tool. Each mode is then further defined to satisfy its operating function.

Figure 5.2   *System control flow chart*

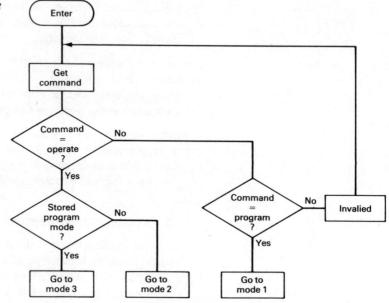

Consider mode 3, 'Drill from stored information'. In this mode the co-ordinates of the required holes are stored in RAM and need to be accessed sequentially by the processor. Figure 5.3 shows a 'first attempt' flow chart to satisfy the function; however, it leaves much to be desired!

### Exercise 5.1

List some of the things that you would need to know to implement the stored co-ordinate mode and then draw a flow chart to illustrate your solution.

Figure 5.4 shows a completed flow-chart for mode 3 operation. However, remember that, as with any problem, your solution may be equally as good; the acid test is the answer to the question 'does it work to the specification?' If the answer is an unqualified yes then either solution is a good one.

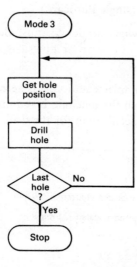

Figure 5.3   *Outline flow chart for mode 3*

Flow charts are now produced for each of the other modes of operation and for initialization and general 'housekeeping'. Now that the problem is becoming somewhat more defined a division between hardware and software can be implemented. This division will be influenced to a large extent by the choice of processor. Each has its strengths and weaknesses for a given application. For general control applications the 8-bit processor is generally adequate. The type of processor will also be influenced by the number of units to be manufactured; for a small number the standard multichip system will be chosen, but if large numbers are to be produced then the single-

Figure 5.4  *Flow chart mode 3*

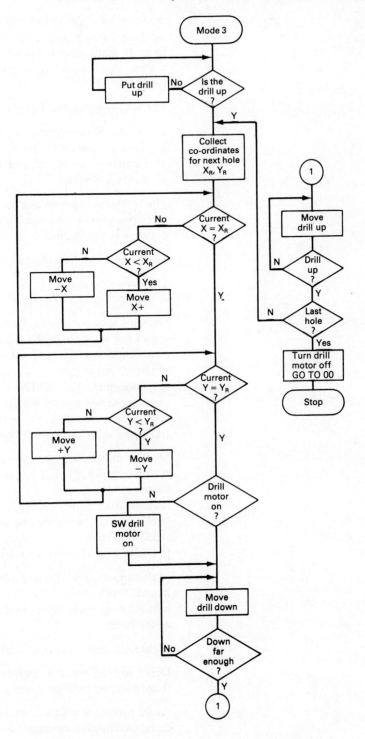

chip processor can cut production costs dramatically, even when the masking costs of the custom ROM have been divided between the units. In many cases for development a multichip system is used with a view to changing to a single-chip system when the product is finalized.

The tool requires the following operations:

1  Worktable movements in $X$ and $Y$ axes.
2  Tool movement in the $Z$ axis.
3  Display of current $X$ and $Y$ positions.
4  Keyboard input.

The mechanical provisions for the tool are to move a worktable to a given position in the $X$ and $Y$ planes and to move the drill head a given distance in the $Z$ plane.

Movements can be obtained in two ways:

1  Servo systems.
2  Stepper motor drives.

Option 1 is essentially a closed-loop system which will necessitate the use of linear or rotary encoders to define the table position at any time. These encoders are expensive if reasonable resolution is to be obtained. An alternative is to use a linear transducer for each axis and an analogue to digital converter (ADC) to provide the microprocessor system with information. This can be either slow in response or complex in operation. Additionally, some means of fixing the table in its rest position is needed when drilling occurs.

Stepper motors are precise, open-loop control devices. Provided a reference co-ordinate is known, a given number of steps input will cause the spindle to rotate a precise number of revolutions (or part thereof) and, provided the table drive unit is free from excessive backlash, this sytem can be used to position the table accurately. Additionally, when stationary but powered up, the stepper motor provides a good brake to hold the table fixed in position.

In this application, 200 step/rev stepper motors are used for both the $X$ and $Y$ axis drives. The $Z$ axis is less critical, and hence the DC motor/linear transducer/ADC method is used to control the depth of the hole.

Figure 5.5 shows the $X$–$Y$ table complete with motors.

Drive for the stepper motors can be either by a dedicated chip (hardware) or by a program-generated sequence.

As the processor will have available time, the table movement function could well be performed by software. In the case of the $Z$ axis the only option is for software control, with the only hardware requirement being power-level interfacing.

Figure 5.5 *The drill with* X–Y *table and motors*

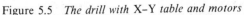

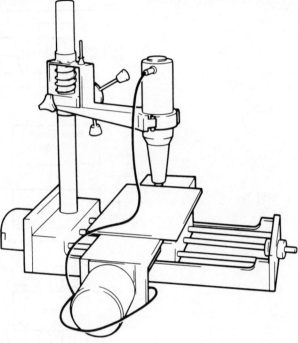

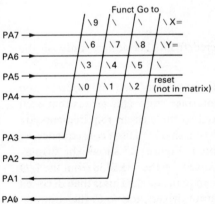

Figure 5.6 *The keyboard connections*

### The man–machine interface

Two further important parts need to be defined. These are the interface between the operator and the machine, i.e. the keyboard and display.

Options are available for both of these. Considering the keyboard, either dedicated encoder logic can be used or, as examined in Chapter 3, a software solution can be found. The latter solution is adopted in this unit to reduce cost and chip count. See Figure 5.6 for the wiring structure to be accepted for a simple keyboard.

For the display, the available options are:

1 Software scan.
2 Dedicated display driver chip.
3 Self-scan display.
4 Latched display.

In case 1 the time overhead of keeping the display refreshed becomes too great and the system operation is too slow. Option 2 is an elegant one in that it allows the display to be multiplexed, so keeping the current requirements down but only requiring updates from the processor when the display is changed.

Option 3 provides similar functions to 2 but includes the display

Figure 5.7   *The display decoding and drivers*

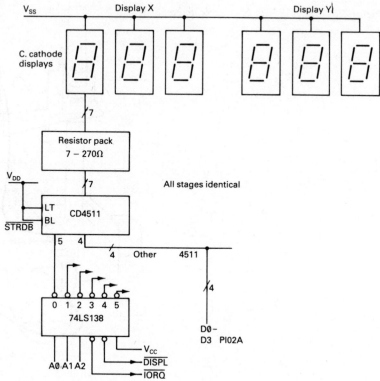

devices. Both of these are rejected from this design study on the grounds of cost.

Option 4 provides each display device with its own I/O 'address' and data is stored in a latch in the interface chip. This means that each display is permanently illuminated so the current requirements are greater. However, for this application that is of little consequence, so this is the solution finally adopted. Figure 5.7 shows the display circuitry used. Each 4511 is 'addressed' by the 74LSI38 from the I/O address bus and BCD data is written to the latch. This is then decoded by the chip and passed to the output drivers and so to the common cathode LED displays. (See Appendix 2 for data sheet).

Having defined the software options, program planning can take place. At the same time, the hardware necessary can be constructed. These parallel operations can reduce project lead time considerably, and full use can be made of specialists in their own areas.

### Selection of the processor

The needs of this system are easily satisfied by an 8-bit processor.

The possible options are:

1   Motorola MC 6800 Series
2   Intel 8085
3   Zilog Z80
4   RCA 1802
5   MOS 6502

In certain cases, constraints will be placed on the selection by 'in-house' standards of the company, but setting these aside there are some very good reasons for not selecting some processors.

One is that of availability. Are the selected processor *and* its peripheral devices available from more than one source? Are the chips still recommended for new designs? These are questions which should be answered before technical limitations are considered. A second area of examination is the manufacturer's back-up and support in the way of documentation and development aids. Most of the popular processors have developments systems available but problems can arise in some cases, so a thorough investigation of engineering support is recommended.

The 8085 can be considered to have an instruction set which is a subset of that used by the Z80, and both are updates of the 8080 CPU which is widely used. However, in real-time applications all these processors have limitations. If the timing diagrams and instruction execution times of these processors are compared with the 6802, then it will become obvious that the Motorola series will execute an instruction in less time than either of the others. Compensation is obtained by operating the Intel and Zilog at higher clock speeds (up to 4 MHz), but this produces memory limitations.

A further point for consideration is how the processor handles interrupts, as for many applications – and for this one in particular – the speed of the response can be critical. The 8085 and the Z80 suffer a serious drawback on interrupt response. Whilst the Z80 has a very flexible interrupt structure, only the program counter is saved in the stack automatically on an interrupt call. This also applies to the 8085. This means that the program must carry out the procedure of saving and returning status within the ISR(s).

The Motorola 6802 etc. automatically saves the entire processor status on an interrupt call. However, only two levels of hardware interrupt are directly available.

For this application, the Zilog Z80 processor was chosen, together with its PIO. 2 Kbytes of ROM (Intel etc. 2716) and 2 Kbytes of CMOS RAM are provided. The latter may be battery backed if required.

Before looking in detail at the hardware it is worth providing a few notes on the Zilog Z80 series of chips. The Z80 CPU is similar in operation to the 8085 previously studied but without the complications of multiplexed data and address buses. The main difference is in the expanded instruction set. The Z80 will run all of the 8080 instructions and most of the 8085 ones except SIM and RIM. Additionally the interrupt structure is different, so the processors are not totally hardware compatible, but for this application the differences will not be inconvenient.

The peripheral device – Z80 PIO – is somewhat different from any device previously examined. A data sheet for the PIO can be found in Appendix 2. As with the Intel PPI etc. the device can be programmed as an input, output or bidirectional port by writing a control word to the control register. However, as the Z80 uses a system of vectored interrupts, the CPU expects the interrupting port to supply the low byte of an address. To achieve this the PIO also contains an interrupt vector register, which must be loaded if the port is to operate under interrupt control. This register is accessed via the control address and is identified by having the LSB (DO) of the control word set to 0. The mode control word has this bit set to a 1. Bits 1 to 7 of the interrupt control word define the low byte of a memory address (lower byte must therefore be 00,02,04...etc to FE, i.e. an *even* numbered address). The upper byte of the memory address is supplied by the I register in the CPU. The address thus specified then holds the *address* of the interrupt service routine for this particular port. So by maintaining a table of ISR addresses in memory each I/O device can be serviced rapidly without the necessity of polling each port in turn until the one requiring service is found.

### System design

Each half of the design (software and processor hardware) requires information from each other to enable them to proceed smoothly. However, the detail of the software routines can be planned and the flow charts produced as the hardware requirements are worked out.

### Outline hardware details

Processor:   Zilog Z80
ROM:        2 Kbytes EPROM type 2718
RAM:        2 Kbytes CMOS   type 6116
Stepper motor control *X*, *Y*, 1 parallel port each
*Z* axis drive:                  2 bits parallel port
*Z* axis motor control:          1 bit parallel port
*Z* axis transducer (via ADC): 8 bits parallel port
*X* and *Y* 'point 00' switches: 2 bits parallel port
Keyboard 4×4 matrix:         2 parallel ports (4 bits only)
Total I/O:                       3×Z80 – P10

As the system is unlikely to be expanded to its full 64 Kbytes of memory, full address decoding is not required so simple logic can be used.

### The hardware

Figure 5.8 shows the circuit diagram of the processor board. Construction could at an early stage be a mixture of wire-wrap and stripboard but a printed circuit board would ease construction and reduce wiring errors.

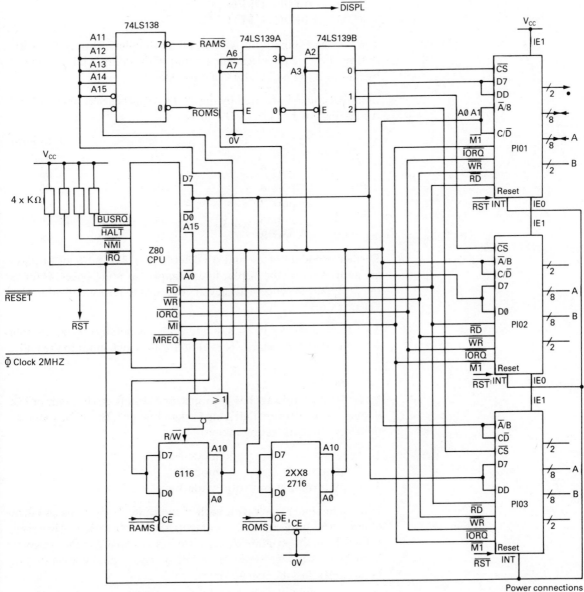

Figure 5.8   *Circuit diagram of the complete processor system*

Memory decoding is accomplished by a single 74LS138 3-to-8 line decoder. Address lines A0 to A10 are connected directly to both memory ICs. The remaining lines – A11 to A15 – are decoded by the 74LS138. A11, 12 and 13 are used as the select inputs whilst A14 and A15 are used as enable signals for the decoder. This ensures that the decoder is only active for addresses with the boundaries 4000–7FFF. Outputs 0 and 6 are used for RAM and ROM select respectively, giving addresses:

RAM     4000H–47FFH
ROM     7000H–77FFH

The I/O decoding is somewhat less than complete. However, the use of three PIOs requires a small amount of decoding. This is achieved by half of a 74LS139 2-to-4 line decoder.

Lines A0 and A1 are used by the PIO as register select lines. A2 and A3 are decoded to give I/O addresses:

|       |        | *Data* | *Control* |
|-------|--------|--------|-----------|
| PIA 1 | port A | 00     | 01        |
|       | port B | 02     | 03        |
| PIA 2 | port A | 04     | 05        |
|       | port B | 06     | 07        |
| PIA 3 | port A | 08     | 09        |
|       | port B | 0A     | 0B        |

Further decoding of A7 and A6 determines if I/O ports or displays are addressed. As the display latches appear as I/O devices, as far as the processor is concerned they are decoded to respond to addresses C0H to C5H.

The clock generator IC7 – a, b and c – is divided by the JK Bistable IC8 to ensure a 1:1 mark:space ratio for the clock input to the CPU.

### The software

Now that the hardware has been specified, a start can be made on the design of the software. Referring back to Figure 5.2 it will be seen that there are three operating modes:

1  Program                   mode 1
2  Find and drill            mode 2
3  Copy (program control)  mode 3

Now is the time to subdivide each of these into its elements and then to write the program which controls the whole system. However, before the main program can operate it is essential that the system is brought to a known state, so an initialization routine must be run before releasing the unit to the operator.

For a project of this size it is essential that the software be modular, i.e. reduced to a number of small routines which can be debugged and tested in isolation, then linked together to produce the complete object code which will reside in the EPROM.

Mode 3 has already been described in Figure 5.4.

Figure 5.9 to 5.11 define the steps in the modes. Note that there are some routines common to each flow chart, i.e. keyboard input, LED output, etc. These can be taken out and defined separately. Each mode can then be charted and programmed along these lines.

Figure 5.9 *Initialization routine flow chart*

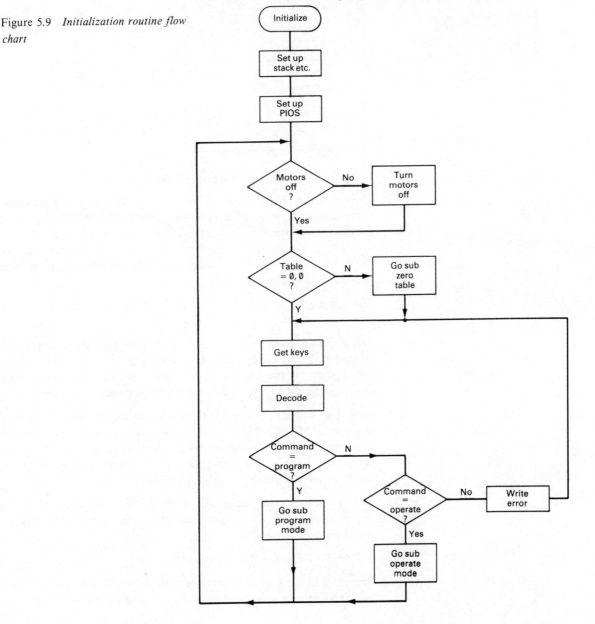

Figure 5.10  *Flow chart – program mode*

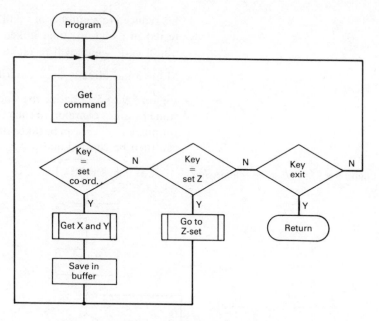

Let us examine each peripheral in turn in more detail.

The keyboard (Figure 5.6) consists of the numbers 0 to 9 plus five command keys and RESET, making a total of 16 keys. The keyboard is software maintained and communicates via one port of a PIO, port A of PIO 2. The use of port A allows the programming of the PIA to be in mode 3 (bit). This programming is performed during the initialization routine.

This arrangement of the keyboard allows the use of the keyboard scan and decode program introduced in Chapter 2.

The display refresh program is the next to be assembled and tested (Figure 5.12). Again, the initialization program is responsible for the initial state of the display buffer. This routine only moves and converts the data from DISBUFF into the display registers.

Next the stepper motor routines are organized. The minimum movement of the table is 0.05 in, so we now need to know the specification of the leadscrew to be used. A 0.5 in AF screw has twenty threads per inch. This gives a pitch of 0.05 in; hence one turn of the screw will cause the table to move 0.05 in.

The $Y$ axis of the table has a length of 160 mm, which is approximately equal to 6.30 in. So to move the table from one end to the other will require:

$$2 \times 6.3 \times 10 \text{ turns} = 126 \text{ turns}$$

Therefore in the $X$ axis the distance is 10 cm, which is nearly 4.0 in. So this axis requires a maximum of eighty turns of the motor.

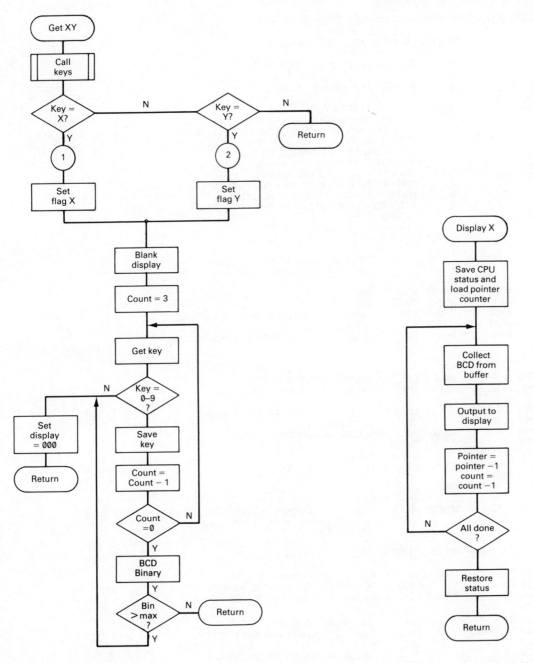

Figure 5.11 *Flow chart – get* XY

Figure 5.12 *Flow chart – display* X Y

The program can be arranged as an inner program which rotates the motor one complete revolution and expects, as a parameter in the D register, the number of revolutions.

The stepper drive program appears in Figure 5.13.

Figure 5.13   *Stepper motor drive program for 200 step/rev motor*

```
Tektronix       Z80 ASM V3.3  STEP DRIVE           Page 1

00002                         ;       STEPPER MOTOR DRIVE SUBROUTINE
00003                         ;
00004                         ;       WRITTEN MAY 1982 P. D. S.
00005                         ;
00006                         ;               DESCRIPTION
00007                         ;               ===========
00008                         ;       THIS PROGRAM OUTPUTS A SEQUENCE OF BITS ON
00009                         ;       PORT A TO DRIVE A 200 STEP/REV. MOTOR.
00010                         ;       THE SEQUENCE IS ARRANGED TO GIVE
00011                         ;       HALF STEP DRIVE, SO ALLOWING
00012                         ;       GREATER PRECISION AND CONTROL.
00013                         ;       PARAMETERS PASSED TO THIS ROUTINE ARE :-
00014                         ;       NUMBER OF REVOLUTIONS IN REGISTER D
00015                         ;       DIRECTION OF ROTATION IN MEMORY LOCATION
00016                         ;         DFLAG
00017                         ;
00018                         ;
00019                         ;
00020                         ;               TABLE OF EQUATES AND CONSTANTS
00021                         ;               ==============================
00022                         ;
00023       0080      ADATA   EQU     80H
00024       0082      ACON    EQU     82H
00025       0300      TIME    EQU     0300H           ;SPEED CONTROL CONSTANT
00026       2F00      NUMRV   EQU     2F00H           ;
00027       2F02      DFLAG   EQU     2F02H           ;0=CLOCKWISE FF=ANTICLOCK.
00028                         ;
00029                         ;
00030                         ;
00031  0000 3E0F      INIT    LD      A,0FH           ;SET UP PORT
00032  0002 D382              OUT     (ACON),A        ;A FOR OUTPUT
00033                         ;
00034  0004 DD21002F          LD      IX,NUMRV        ;GET NUMBER OF REVS.
00035  0008 DD5600            LD      D,(IX+0)
00036  000B 3A022F    WCHWA   LD      A,(DFLAG)       ;GET DIRECTION
00037  000E B7                OR      A               ;SET FLAGS
00038  000F F21700 >          JP      P,FWD           ;
00039  0012 214000 >          LD      HL,DOWN         ;REVERSE DIRECTION
00040  0015 1805              JR      WUNREV          ;RUN MOTOR
00041  0017 214800 > FWD      LD      HL,UP           ;FORWARDS
00042  001A 1800              JR      WUNREV          ;RUN MOTOR
00043  001C 0E32     WUNREV   LD      C,50            ;NUMBER OF LOOPS OF PROGRAM
00044                         ;               200 STEPS/REV, 4 STEPS PER LOOP =
                                                      200/4 = 50


Tektronix       Z80 ASM V3.3  STEP DRIVE           Page 2

00046  001E 0608      ONEST   LD      B,8             ;NUMBER OF STEPS
00047  0020 E5                PUSH    HL              ;SAVE TABLE POINTER
00048  0021 7E       RUNLUP   LD      A,(HL)          ;SET FIRST STEP

                                                                  PAGE002
00049  0022 D380              OUT     (ADATA),A       ;AND MOVE MOTOR
00050  0024 CD3400 >          CALL    DWELL           ;WAIT FOR MOTOR TO CATCH UP
00051  0027 23                INC     HL              ;GET READY FOR NEXT STEP
00052  0028 10F7              DJNZ    RUNLUP          ;AND PROCEED IF NOT 8
00053  002A E1                POP     HL              ;
00054  002B 18F1              JR      ONEST           ;DO NEXT FOUR STEPS
00055  002D 0D                DEC     C               ;ALL DONE?
00056  002E 20EE              JR      NZ,ONEST
00057  0030 15                DEC     D ;CHECK IF ALL STEPS DONE
00058  0031 20D8              JR      NZ,WCHWA        ;
00059  0033 C9                RET                     ;RETURN TO CALLING PROGRAM
00060                         ;
00061                         ;       DELAY TO SET MOTOR SPEED
```

*continued*

```
00062                   ;       CHANGE THE CONSTANT -TIME-
00063                   ;       TO CHANGE THE SPEED.
00064                   ;
00065 0034 E5    DWELL   PUSH    HL          ;SAVE POINTER REGISTER
00066 0035 11FFFF        LD      DE,-1       ;DECREMENTER
00067 0038 210003        LD      HL,TIME     ;DELAY TIME
00068 003B 19    WLUP    ADD     HL,DE       ;DECREMENT HL
00069 003C 38FD          JR      C,WLUP      ;UNTIL ZERO
00070 003E E1            POP     HL          ;THEN RESTORE POINTER REG.
00071 003F C9            RET                 ;AND RETURN
00072                   ;
00073                   ;       LOOK UP TABLE OF STEPS
00074                   ;       ======================
00075                   ;
00076 0040 0109080C DOWN BYTE   1,9,8,12,4,6,2,3
00077 0044 04060203
00078 0048 03020604 UP   BYTE   3,2,6,4,12,8,9,1
00079 004C 0C080901
00080                   END
```

```
Tektronix  Z80 ASM V3.3 Symbol Table            Page 3


Scalars


ACON---0082  ADATA--0080  DFLAG--2F02  NUMRV--2F00  TIME---0300

% (default) Section (0050)

DOWN---0040  DWELL--0034  FWD----0017  INIT---0000  ONEST--001E
RUNLUP-0021  UP-----0048  WCHWA--0008  WLUP---003B  WUNREV-001C


79 Source Lines   79 Assembled Lines   45563 Bytes available

            No assembly errors detected
```

The $Z$ axis routine consists of two parts, a motor up–down driver and an ADC subroutine to find the position of the drill head. This routine uses a constant entered under the SET 2 routine, which defines the difference between the maximum hole depth and the position of the drill head at the time the reset button is operated.

With the hardware and most of the software defined, a start can be made on the assembly of each.

*Summary of required software routines*
Keyboard scan and decode
Display refresh
Stepper motor drive
A–D conversion
Command decode
Axis movement – $X$ or $Y$
Axis control – $Z$
Initialization.

*Summary of hardware*
Board size – double Eurocard
CPU – Z80A
I/O – 3 × Z80A PIO
Separate keyboard
Separate display unit
*X* and *Y* stepper motor drivers
d.c. motor controller
ADC (on CPU card)
Linear transducer (LIN slider port, 10 kΩ)
Power supply: 5 V, 5 A; 12 V, 25 A.

*Summary of mechanical requirements*
d.c. minidrill – precision petite or similar
Drill press (stand) to match above
X–Y table (for details see Chapter 6)
Housing for electronic units
Supply of drills for testing
Supply of boards for testing.

### Assembly and test

Now that the requirements have been set out a start can be made on assembling the hardware and software so that they can be tested together. If full in-circuit emulation facilities are available, then it is best to get the hardware well advanced so that the software can be tested with it. If only a more limited development system is available then it is advisable to test routines completely independently first. For example, it is possible to test the stepper motor drive subroutine to see that it does turn the shaft the required number of degrees for a given input parameter. Does the display produce the right numbers when they are preloaded into a buffer and then read out? In this way one can be fairly confident of certain parts of the program before the whole system refuses to work when completely assembled!

It is very easy with a project of this type to 'add bits on', that is, extend the system beyond the original specification. It is necessary to be aware at all times of the original specification, and to complete the project when that specification has been met.

Additionally full and thorough documentation is a *must* with an undertaking of this size. Diagrams, programs and descriptions should be carefully recorded and *dated* so that the last version is easily recognized. Communication is essential especially if more than one person is working on the same project.

### 5.3   Data logging system

The second project provides for collecting analogue data from a number of transducers and formatting it for onward transmission via a serial data link which may be via a modem connected to the public switched telephone network (PSTN).

*Job specification*

1   The system should, each hour, be able to:
   (a)   Collect the state of the following analogue inputs:
      (i)   Air temperature
      (ii)   Atmospheric pressure
      (iii)  Humidity
   (b)   Maintain a wind speed and direction input.
   (c)   Log rainfall at the site over a period of twelve hours.

2   The equipment is to be installed at a site remote from the data monitoring point. Hence the collected information must be formatted for serial data transmission at 110 or 300 bauds either over a private wire connection or via a British Telecom modem and telephone channel.

This section is intended as a series of notes and suggestions to allow the student to develop his own programs and expand the ideas presented. Some programs are presented where new topics are encountered.

**Power supply and consumption**

This remote monitoring and telemetry system produces some unique problems for solution. One possibility is that the mains power supply is likely to be unreliable or, in some cases, unobtainable. Hence provision must be made for battery operation either on a 'float' charge system from the mains or from a natural source of energy such as solar cells or a wind generator.

In either of these two latter cases the input to the battery may well be sporadic, so special attention will have to be paid to the power consumption. CMOS circuits have a lower power consumption figure than the normal NMOS processors, and are therefore worthy of consideration. The problem with CMOS processors is that the choice is limited. The RCA 1802 or the CMOS MC 6802 is about the only choice at the time of writing. The 6802 recommends itself as the software is available for the normal NMOS processor. For this application, the 6802 CMOS version was chosen.

*Peripherals and data handling*

The choice of peripherals and handling of data must now be settled.

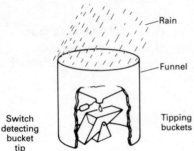

Figure 5.14   *Rainfall measuring unit*

Figure 5.15   *A slotted opto-switch RS type 306-061*

The 6802 has limited interrupt capability. and as a real-time clock is required this will take up one level of interrupt (non-maskable). This leaves the maskable interrupt available for one other interrupt input. Measurement of pressure, temperature, humidity and wind direction are analogue inputs and hence need to be interfaced via an ADC. Wind speed and rainfall, on the other hand, are handled digitally.

### Rainfall

Figure 5.14 shows the device used to measure rainfall. The funnel is 12 cm in diameter. If 1 mm of rain falls over this area, then there will be:

$$\pi 6^2 \times 0.1 \text{ ml} \simeq 11.3 \text{ ml}$$
of water collected.

The design of the 'bucket' is such that when it holds this amount of water it overbalances, empties and presents the other bucket for filling. Each time the bucket tips an impulse is generated by the slotted optoswitch which increments a counter.

Dimensions of the bucket etc. are given in Chapter 6.

### Wind speed

Wind speed is recorded by an external hardware counter. Again, a slotted optoswitch (Figure 5.15) is used to produce an output, but this time the signal is used to increment a 4040 CMOS counter which is interrogated by the processor each minute and reset after reading. The program then converts the number of pulses per minute into wind speed. This counter interfaces to the system via one port of a PIA. Control bit 2 is used to reset the counter to zero at each reading.

### Temperature and humidity

Temperature and humidity measurements are based on the LM 3911 temperature sensor IC from National Semiconductor, although any similar device producing a linear change in output voltage for change in temperature could be used.

The humidity measurement is based on the 'wet and dry bulb' thermometer method, and reference to tables will be necessary for conversion to the correct humidity values (see Appendix 2). Figure 5.16 shows the circuit arrangement of one of the sensors.

### Pressure

Pressure is measured by a National Semiconductor pressure transducer. Once again, this has an output voltage which is proportional to absolute pressure. The circuit is shown in Figure 5.17. Some non-linearity due to temperature changes in the semi-

Figure 5.16 *Connections of the LM3911 temperature sensor*

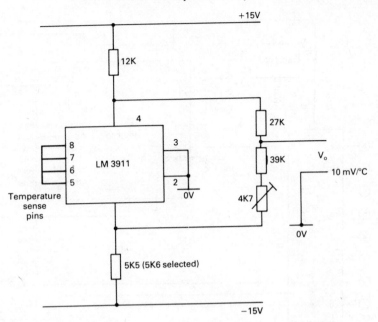

Figure 5.17 *Connecting the LX0503A pressure transducer*

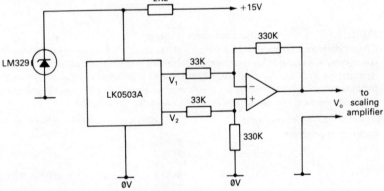

conductor material may be noticed, but these can easily be corrected by the software.

### ADC

As each of the transducers is producing output voltages in the same range, a possibility for A–D conversion is to use a digital voltmeter chip as the input to the system and to multiplex each transducer to the input. The circuit, Figure 5.18, is adapted from a Motorola application note no. AN770, and the program is modified to run under polling control rather than interrupt (Figure 5.19).

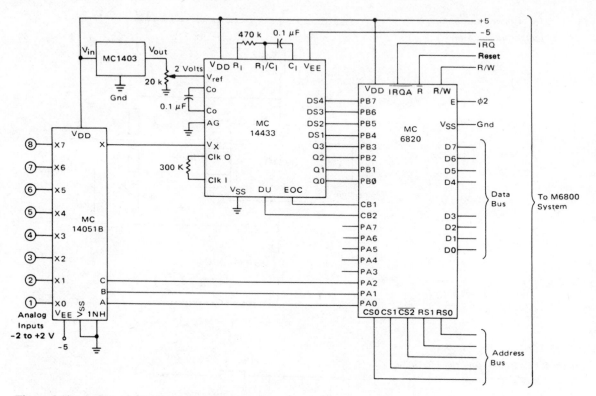

Figure 5.18   *8-channel data acquisition hardware*
*Circuit diagrams external to Motorola products are included as a means of illustrating typical semiconductor applications; consequently, complete information sufficient for construction purposes is not necessarily given. The information in this Application Note has been carefully checked and is believed to be entirely reliable. However, no responsibility is assumed for inaccuracies. Furthermore, such information does not convey to the purchaser of the semiconductor devices described any licence under the patent rights of Motorola Inc, or others.*

Figure 5.19   *8-channel data*
*acquisition program*

```
TEKTRONIX      M6800 ASM V3.3  8. CH. CON.

00002                      ;
00003                      ;8-CHANNEL DATA COLLECTION PROGRAM
00004                      ;
00005                      ;DATE DEPT 1982
00006                      ;
00007                      ;EACH CHANNEL IS STORED AS A B. C. D. NUMBER
00008                      ;WITH THE M. S. BIT OF THE M. S. DIGIT INDICATING
00009                      ;THE POLARITY OF THE SIGNAL, AN F1 IN THE M. S. 1
00010                      ;INDICATES OVER RANGE
00011                      ;
00012                      ;
00013      0050            STORL  EQU 50H          ;TEMP STORE FOR 1X REGIST
00014      0052            POINTER EQU 52H
00015      0054            TEST    EQU 54H
00016                      ;
00017                      ;DATA VALUES STORED 0060H - 007FH
00018      8010      >             ORG 8010H        ;PIA REGISTERS
00019                      ;
00020 8010 01             PIA1AD BYTE 1
00021 8011 01             PIA1AC BYTE 1
00022 8012 01             PIA1BD BYTE 1
00023 8013 01             PIA1BC BYTE 1
00024                      ;
```
*continued*

```
00025                       ;
00026      F800    >    ORG 0F800H
00027                       ;
00028                       ;SET UP PIA AND CONSTANTS
00029                       ;
00030 F800 7F0054           CLR TEST
00031 F803 7F8013 >         CLR PIA1BC        ;SET UP PIA S
00032 F806 7F8012 >         CLR PIA1BD        ;SET FOR INPUT
00033 F809 7F8011 >         CLR PIA1AC
00034 F80C 86FF             LDA A #0FFH       ;SET PORT B FOR OUTPUT
00035 F80E B78010 >         STA A PIA1AD      ;
00036 F811 8634             LDA A #34H        ;
00037 F813 B78013 >         STA A PIA1BC      ;PIA S NOW SET UP
00038 F816 B78011 >         STA A PIA1AC      ;
00039                       ;
00040 F819 8D03             BSR CONV          ;GET DATA
00041 F81B 7EE0D0           JMP 0E0D0H        ;RETURN TO MAIN PROGRAM
00042                       ;
00043                       ;CONVERSION ROUTINES
00044                       ;
00045 F81E CE005C    CONV   LDX #005CH        ;DATA BUFFER - 4
00046 F821 DF50             STX STORL         ;SAVE IX
00047 F823 C604             LDA B #4          ;NUMBER OF DIGITS
00048 F825 D754             STA B TEST        ;
00049 F827 B68012 >         LDA A PIA1BD      ;CLEAR FLAGS
00050 F82A 8637             LDA A #37H        ;
00051 F82C B78013 >         STA A PIA1BC      ;START A CONVERSION
00052 F82F BDF8A2 >         JSR WAIT          ;DUMMY RUN TO SYNC CONVERTER
00053 F832 C607             LDA B #7          ;NUMBER OF CHANNELS-1
00054 F834 B78010 >  N      STA A PIA1AD      ;SET CHAN 7
00055 F837 8602             LDA A #2          ;
00056 F839 9754             STA A TEST        ;
00057 F83B 8637             LDA A #37H        ;START CONVERSION
00058 F83D B78013 >         STA A PIA1BC      ;
00059 F840 BDF8A2 >         JSR WAIT          ;WAIT FOR CONVERSION
00060 F843 8D09             BSR ISR           ;GET CONVERTED DATA
00061 F845 BDF8A2 >         JSR WAIT          ;
00062 F848 8D04             BSR ISR           ;
00063 F84A 5A               DEC B             ;NEXT CHAN
00064 F84B 2AE7             BPL N             ;
00065 F84D 39               RTS               ;ALL DONE
00066                       ;
00067                       ;
00068 F84E 863F     ISR     LDA A #3FH        ;RESET CONVERT COMMAND
00069 F850 B78013 >         STA A PIA1BC      ;
00070 F853 740054           LSR TEST          ;FIRST TIME ROUND?
00071 F856 2446             BCC FIRST         ; YES SO IGNORE
00072 F858 8634             LDA A #34H        ;
00073 F85A B78013 >         STA A PIA1BC      ;
00074 F85D 8610     BEGIN   LDA A #10H        ;
00075 F85F 9752             STA A POINTR      ;
00076 F861 DE50             LDX STORL         ;GET DATA ADDRESS
00077 F863 B68012 > NEXT    LDA A PIA1BD      ;GET CONVERTED VALUE
00078 F866 760054           ROR TEST          ;NEXT DIGIT
00079 F869 DB54             ADD B TEST        ;
00080 F86B 16               TAB               ;SAVE IT
00081 F86C 9452             AND A POINTR      ;ALL DONE?
00082 F86E 27F3             BEQ NEXT          ;YES
00083 F870 780052           ASL POINTR        ;
00084 F873 C40F             AND B #0FH        ;
00085 F875 E704             STA B 4,X         ;SAVE DATA
00086 F877 08               INX               ;NEXT DIGIT
00087 F878 24E9             BCC NEXT          ;
00088 F87A A600             LDA A 0,X         ;CHECK FOR OVER RANGE
00089 F87C 16               TAB               ;TEMP STORE
00090 F87D 840B             AND A #11         ;MASK UPPER PART
00091 F87F 8103             CMP A #3          ;DIGIT = 3?
00092 F881 2714             BEQ OVRNG         ;YES
00093 F883 7F0054           CLR TEST          ;NO ALL O.K.
```

*continued*

```
00094 F886 C40C            AND B #12        ;ALIGN HALF DIGIT
00095 F888 54              LSR B
00096 F889 54              LSR B
00097 F88A 54              LSR B
00098 F88B 760054          ROR TEST         ;PREPARE FOR NEXT
00099 F88E DB54            ADD B TEST       ;
00100 F890 53              COM B            ;INVERT
00101 F891 C481            AND B #81H       ;
00102 F893 E700            STA B 0,X        ;RESTORE
00103 F895 2004            BRA FINE         ;CHANNEL DONE
00104 F897 86F1      OVRNG  LDA A #0F1H      ;ERROR FLAG
00105 F899 A700            STA A 0,X        ;INTO M. S. D.
00106 F89B DF50      FINE   STX STORL        ;UPDATE IX STORE
00107 F89D 39              RTS              ;RETURN
00108 F89E B68012  > FIRST  LDA A PIA1BD     ;CLEAR FLAG
00109 F8A1 39              RTS              ;AND RETURN
00110 F8A2 36        WAIT   PSH A            ;SAVE ACC
00111 F8A3 B68013  > LUK    LDA A PIA1BC     ;GET STATUS
00112 F8A6 2AFB            BPL LUK          ;WAIT FOR CONVERSION
00113 F8A8 32              PUL A            ;
00114 F8A9 39              RTS              ;READY
00115                      END
```

ektronix    M6800 ASM V3.3 Symbol Table
Scalars

  POINTR - 0052          STORL -- 0050          TEST --- 0054

%BKCONO (default) Section (F8AA)

```
BEGIN -- F85D    CONV --- F81E    FINE --- F89B    FIRST --- F89E
ISR ---- F84E    LUK ---- F8A3    N ------ F834    NEXT ---- F863
OVRNG -- F897    PIA1AC - 8011    PIA1AD - 8010    PIA1BC -- 8013
PIA1BD - 8012    WAIT --- F8A2
```

115 Source Lines    115 Assembled Lines    47168 Bytes available

         No assembly errors detected

The use of a DVM chip provides a further advantage in that the output is in BCD form rather than pure binary. So, by appropriate scaling, it is possible to obtain a direct reading of the measured quantity.

*Clock*

As the system has to provide information each hour, it is necessary to provide a clock of some sort on the system. A number of solutions are possible:

1  Provide a dedicated clock chip on the bus.
2  Provide a 'software clock' under interrupt control.
3  Provide a 'radio code' clock receiving the MSF transmissions.

For a project of this nature the last possibility is worthy of consideration, especially if supply interruptions are to be experienced, as the clock time is transmitted as a binary code each minute. Therefore, on power-up after failure, the maximum time the processor does not have the correct time will be less than 1 minute. The disadvantage is that more complex hardware will be required.

Solution 1, whilst giving the time directly, would require that a separate crystal would be needed to drive the clock from a battery supply under power fail conditions or provision be made to recalibrate the clock remotely.

Solution 2 is the most cost-effective to implement. The processor system already has an accurate crystal for its clock. This can be counted down and used to provide an interupt every 20 ms or 1 s. An interrupt service routine then maintains a time buffer in memory. If the whole processor system has a float battery, the problems of power fail will not corrupt the timekeeping of the system.

### System completion

The system has to provide its information on a telephone link. Therefore parallel to serial conversion is carried out using a UART-type device, i.e. the ACIA (6850) with modern control signals RTS-CTS, etc. as required. This also provides a convenient way of setting up the system initially. If, as well as the modem connection, software is provided to drive a portable terminal such as the silent 700 from Texas Instruments, then the I/O problem is solved for initialization.

Figure 5.20  *Block diagram of the remote monitoring system*

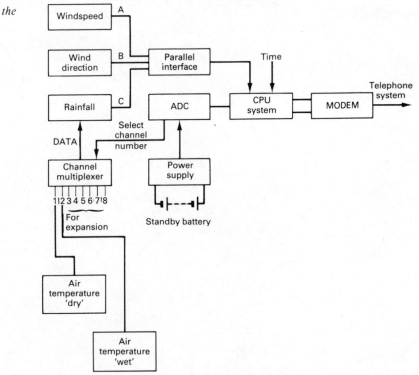

Figure 5.20 shows an overview of a suggested system to implement this project. However, it should not be taken that this is the *only* solution. Other methods and devices for measurement will suggest themselves to the student; in addition new devices will appear on the market after this text is published.

# Chapter 6  Construction techniques

These notes are intended as a guide to those who wish to duplicate the hardware described in previous chapters. At the end of this chapter there is a list of firms which have supplied parts during the development of the units described.

If the reader decides to build further expansion units for the HEKTOR computer it should be remembered that there is no buffering of the processor system buses, and so only a single load or so should be added if bus buffers/transceivers are not provided. The units described previously are of the former type and no problems have been experienced during development.

## 6.1  Circuit construction

The method of construction of individual peripheral boards for experiments will largely be dictated by what is available, but wherever possible a typical 'industrial' technique should be chosen.

It is assumed that students working on level IV and V programmes will have gained a degree of skill in circuit construction and, possibly, printed circuit manufacture and assembly.

The peripheral boards for HEKTOR described in previous chapters were mostly constructed on a standard size strip (vero) board available from RS Components Ltd Stock No. 434-217. While this board is rather large for some units it was considered best to keep to a single size.

The edge connector on HEKTOR is double sided, and hence conversion to a single-sided board is necessary while keeping cost within reasonable figures. To this end a short extension lead was made up with a mating edge connector at one end (McMurdo EC100 type from Farnell Electronic Components) and 0.1 in Molex plugs (type KK) on the other. The lead serves all boards, the edge connector end being left connected at all times. It is important that the length of the ribbon cable does not exceed about 10–15 cm otherwise the lead capacity will have an effect on the signals.

The method of circuit construction using stripboard is probably well known to the reader, but some time can be saved by using 22 gauge solder-through enamel-covered wire to make the necessary intertrack connections, instead of the more usual PVC covered wire.

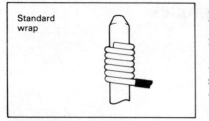

Standard wrap – only the bare wire is wrapped around the terminal. Normally preferred with large gauge wire.

Figure 6.1   *Regular or standard wrap connection*

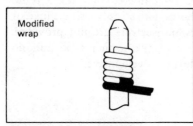

Modified wrap – The first half to two turns are made with insulation wrapped around the terminal. These turns are in addition to the recommended turns made with bare wire.

This type of wrap greatly increases the ability of the connection to withstand vibration, and also reduces wire breakage.

Figure 6.2   *Modified wrap connection*

As an alternative to stripboard, wire-wrap construction may be used. This method has gained popularity of late and provides a good reliable method of interconnection.

A wire-wrapped connection is made by winding the wire around the sharp corners of a terminal post under tension. The bending of the wire around the sharp edge breaks down the surface oxide layers and as the metals are crushed a low-resistance, gas-tight joint is produced. Solderless wrapped connections have excellent mechanical and electrical stability owing to this gas-tight contact area. They remain stable through exposure to severe temperature changes, humidity, corrosive atmospheres and vibration. The high shearing force of the wire at the corner of the terminal and high contact pressure produces an intimate clean oxide-free metal-to-metal contact with a large contact area.

Two main types of wrap are used:

1   Regular or standard wrap (see Figure 6.1).
2   Modified wrap (see Figure 6.2).

For the type of work considered here the modified wrap is to be preferred, as this gives extra mechanical stability to the connection.

Solid wire is used for wire-wrap connections and the most popular gauge is 30 s.w.g. or 0.25 mm.

A distinct advantage of wire-wrapping is the ease with which a wire can be removed or replaced for modification. An unwrap tool slipped over the terminal and then rotated removes the unwanted wire in seconds without damage to the terminal or component.

A range of tools are available for wire-wrapping connections. Figure 6.3 illustrates a typical battery-operated wrapping tool, and Figure 6.4 shows a tool which will cut and strip the insulation for that wire to the correct length for insertion into the wrapping tool.

Figure 6.3   *Wire-wrap tool*

Figure 6.4   *Wire stripper for wrapping*

Figure 6.5 *Breadboard*

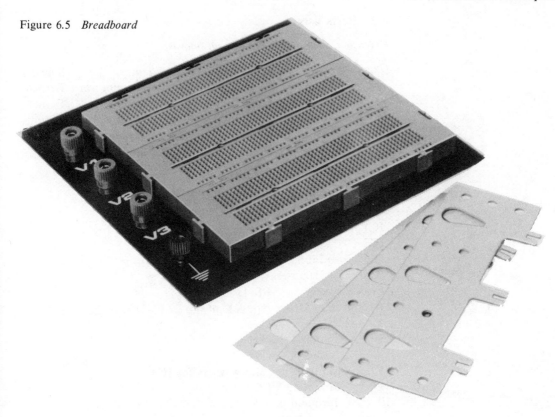

The other alternative for trying out circuits is to use a prototyping board or breadboard. These are available from many suppliers under brand names, and one is illustrated in Figure 6.5.

Component leads and wire are pushed into the holes and interconnection is made through conductors inside the board. These devices are useful for trying out a circuit with a few components, but they can get messy and unreliable as the number of components in a circuit increases.

No mention has been made of printed circuits for these projects. The time required to produce a printed circuit for one of the boards described is somewhat excessive when compared with the time taken to produce the same unit using other techniques. However, it is hoped that PCBs may become available for some of the items.

The use of IC sockets is recommended *with one proviso* – that is, use only the best quality, preferably the turned-pin type of socket, as there is nothing worse than spending many hours tracing a fault which in the end is caused by an intermittent pin in an IC socket.

## 6.2 Parts lists

### Extension cable (HEKTOR to all boards)

| Quantity | Manufacturer's pt no. | Description | Supplier |
|---|---|---|---|
| 4 | 22–01–2105 | Molex KK connector | Farnell |
| 40 | 08–50–0136 | Crimp terminal | Farnell |
| 30 cm | | 40-way ribbon cable | Various |
| 1 | EC100–4040–S–SG | Edge connector | Farnell |

### Figures 2.7 and 2.8

| C1 | 220 nF Mullard C230 |
|---|---|
| C2 | 470 nF 35 V tantalum bead |
| IC1 | I8212 |
| IC2 | 74LS02 |
| IC3 | 74LS138 |
| IC4 | 7805 |

Circuit board      RS 434-217
4 connectors Molex type KK pt no. 22-03-2101 or 22-27-2101

| 1 | 24 pin IC socket |
|---|---|
| 1 | 16 pin IC socket |
| 1 | 14 pin IC socket |
| 1 | M3 bolt, nut and washers for IC4 |
| 1 | Mounting kit for IC4 |
| 10 | Terminal pins RS no. 433-624 or equivalent |

### Figure 2.9

| 8 | 270 $\Omega$ 0.25 W c.f. resistors |
|---|---|
| 8 | 1 k$\Omega$ 0.25 W c.f. resistors |
| 8 | LEDs TIL 209 or similar |
| 8 | BC107/8 or equivalent. |

This is built as an addition to Figure 2.8.

### Figure 2.11

| Q1 | BC107 | Q2 | BFY51 |
|---|---|---|---|
| Q3 | TIP121 | Q4 | TIP121 |
| Q5 | TIP121 | Q6 | TIP121 |
| Q7 | BFY51 | Q8 | BC107 |

| R2,R3 | 10 k$\Omega$ 0.25 W |
|---|---|
| R1,R4 | 47 k$\Omega$ 0.25 W |

| 10 | Terminal pins RS type 433-624 |
|---|---|
| 1 | Circuit board RS type 434-217 |
| 4 | Heat sinks type TV4 (electrovalue) |

*Figure 2.13*

| C1 | 220 nF Mullard C280 type |
| C2 | 470 nF 35 V tantalum bead |

| IC1 | I8255 |
| IC2 | 74LS138 |
| IC3 | 7805 |

| 1 | Circuit board RS 433-624 |
| 30 | Terminal pins RS 433-624 |
| 4 | Molex connectors type KK pt no. 22-03-2101 or 22-27-2101 |
| 1 | 40 pin IC socket |
| 1 | 16 pin IC socket |
| 1 | Heatsink type TV4 |

| 24 | Resisitors apprx. 10 kΩ 0.25 W for pull-ups on unused peripheral inputs. |

*Figure 2.16*

| R1, R2, R3 | 1kΩ 0.25 W |
| S1, S2, S3 | Keyboard switch RS type 337-611 |

This circuit is implemented in a prototype board.

*Figure 2.19*

| R1, R2, R3 | 10 kΩ 0.25 W |

| 9 | Key switches RS 337-611 or similar |
| 1 | Circuit board RS 434-217 |
| 10 | Terminal pins RS 433-624 or similar |
| 10 cm | Ribbon cable 10 Way. |

*Figure 2.22*

| R1, R2, R3 | 1 kΩ 0.25 W |
| R4–R10 | 270 Ω 0.25 W |

| IC1 | ULN2003 |

| D1, D2, D3 | DL707 |

| Q1, Q2, Q3 | BC107 |

Circuit board RS 434-217
Terminal pins RS 433-624 15 off

| 3 | IC sockets 14 pin |
| 1 | IC sockets 16 pin |

*Figure 2.24*

Note: all resistors are 2%, 0.25 W unless noted

| | |
|---|---|
| R1 | 220 Ω 5% |
| R2 | 10 kΩ |
| R3 | 2 × 10 kΩ in series |
| R4 | 3 × 120 kΩ in parallel |
| R5 | 3 × 240 kΩ in parallel |

| | |
|---|---|
| D1 | BZY88C4V7 |
| IC1 | 4066 (Quad CMOS switch) |
| IC2 | CA3130 |

| | |
|---|---|
| 1 | Circuit board |
| 1 | IC socket 14 pin |
| 1 | IC socket  8 pin |

Terminal pins as required.

*Figure 2.26*

| | |
|---|---|
| R1 | 4.7 kΩ 0.25 W 5% |
| R2 | 6.8 kΩ 0.25 W 5% |
| R3 | 18 kΩ  0.25 W 5% |
| R4 | 10 kΩ  0.25 W pre-set potentiometer |

| | |
|---|---|
| C1 | 220 nF 100 V |
| C2 | 100 pF C632 ceramic |

| | |
|---|---|
| IC1 | ZN425E |
| IC2 | CA3140 |

Plus usual board, pins, sockets etc.

*Figure 2.30*

All resistors 0.15 W, 5%

| | |
|---|---|
| R1 | 1 kΩ |
| R2 | 15 kΩ |
| R3 | 100 kΩ ten-turn preset |
| R4 | 4.7 kΩ |
| R5 | 4.7 kΩ |

| | |
|---|---|
| C1 | 220 nF 100 V plastic film |
| C2 | 56 pF min. ceramic |
| C3 | 1 nF polystyrene |

| | |
|---|---|
| IC1 | ZN425E |
| IC2 | CA3130 |
| IC3 | 7400 |
| IC4 | NE7555 |

Plus board and terminal pins etc.

*Figure 2.43*

| R1, R2, R3 | 10 kΩ 0.25 W |
|---|---|
| D1, D2, D3 | 1n4148 |
| Q1, Q3, Q5 | RC107 |
| Q2, Q4, Q6 | BFY50 |

Circuit board etc.

*Figure 2.44*

| IC1 | I8251 |
|---|---|
| IC2 | 74LS138 |
| IC3 | NE555 |
| IC4 | 7805 |
| C1 | 220 nF 100 V polyester |
| C2 | 470 nF 35 V tantalum |
| C3 | 100 nF 50 V disc ceramic |
| C4 | 47 nF 5% polystyrene |
| R1 | 4.7 kΩ 0.25 W |
| R2 | 1 kΩ 0.25 W |
| R3 | 10 kΩ preset potentiometer |
| 1 | 28-pin IC socket |
| 1 | 16-pin IC socket |
| 1 | 8-pin IC socket |

Board, pins etc.

*Figure 3.4*

| D1, D2 | 1N4002 |
|---|---|
| R1 | 1 kΩ 5% 0.25 W |
| R2 | 100 kΩ 5% 0.25 W |
| IC1 | 7404 |

26 V 2% regulation power supply

1       24-pin zero insertion force (ZIF) IC socket

This unit is built on to a 8255 peripheral board and forms the complete hardware for programming 2716-type EPROMS. If it is required to program 2732/2764 types then the remaining bits of port C can be used for the extra address lines.

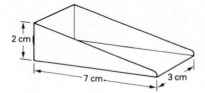

Figure 6.6    *Rainfall bucket detail*

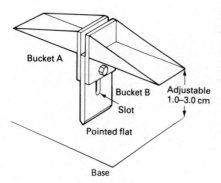

Figure 6.7    *Rainfall bucket adjustment*

## 6.3    Rainfall bucket details

The bucket is constructed from aluminium sheet and coated with an epoxy resin (Araldite or similar) to provide waterproofing. Dimensions are shown in Figure 6.6. It is free standing within an enclosure. As calculated in the text the bucket should tip when 11.3 ml of rain has been collected. To this end the height of the bucket and hence its point of balance are made adjustable as shown in Figure 6.7.

The bucket can then be 'calibrated' by introducing water from a measuring tube and then adjusting the height of the fulcrum until it overbalances with the exact amount of water. It is suggested that there will not be a very large inaccuracy if the bucket tips at 11 ml or 11.5 ml instead of 11.3 ml, as there will always be some loss by evaporation during periods of low rainfall.

The student is directed to the leaflets published by the Meteorological Office on weather forecasting as well as some of the better physics texts for details of the conditions under which parameters of this project are measured.

## 6.4    List of suppliers

This list of suppliers, whilst not exhaustive, includes those that have been found to maintain stocks of all components used in the exercises and projects described in previous chapters.

Enquiries to individual companies should obviously be accompanied with a stamped, self-addressed envelope for a reply.

Electrovalue Ltd
28 St. Judas Road
Englefield Green
Egham
Surrey TW20 0HB

Capacitors, resistors, ICs
(TTL, CMOS and some LSI),
data books, tools, LEDs etc.

Farnell Electronic
   Components Ltd
Canal Road
Leeds LS12 2TU

Broad line supplier (almost
everything)

Happy Memories
Gladestry
Kington
Herefordshire HR5 3NY

ICs

| | |
|---|---|
| Midwich Computer Co Ltd<br>Rickinghall House<br>Hinderelay Road<br>Rickinghall<br>Suffolk IP22 1HH<br>Tel. 0379 898751 | ICs, IC stockists, data hardware,<br>clock crystals, keyboards etc. |
| P.N.P. Communications<br>62 Lawes Avenue<br>Newhaven<br>East Sussex BN9 9SB<br>Tel. 0273 574465 | ICs, sockets, PCBs,<br>connectors, kits |
| R.S. Components Ltd<br>P.O. Box 427<br>13–17, Epworth Street<br>London EC2P 2HA | Broad range supplier |
| Stewart of Reading<br>110 Wykeham Road<br>Reading<br>Berks RG6 1PL<br>Tel. 0734 68041 | Stepper motors, transformers,<br>equipment cases, teletypes etc. |
| Technomatic Ltd<br>17 Burnley Road<br>London NW10 1ED | ICs, IC sockets,<br>connectors, switches etc. |
| Verospeed Ltd<br>Stanstead Road<br>Bovatt Wood<br>Eastleigh<br>Hants S/5 4ZY | Boards, cases, tools |

These addresses are correct at the time of compiling this list, however, no guarantee is given as to their current accuracy.

# Appendix 1 Using a printer with HEKTOR

The serial I/O port on the system supports an RS 232 (V24) interface for hard copy. No firmware is provided for interfacing a keyboard via this port but output is obtainable by typing CONTROL and P together instead of RETURN as an input terminator. In response to CONTROL P the system asks you to SET PRINTER by typing RETURN then outputs the specified text on to the printer.

### Interfacing a printer

Figure A1.1 shows the connection details for a printer connected to the serial interface connector SK6.

Figure A1.1   *Printer connection details (source: Open University HEKTOR User Manual)*

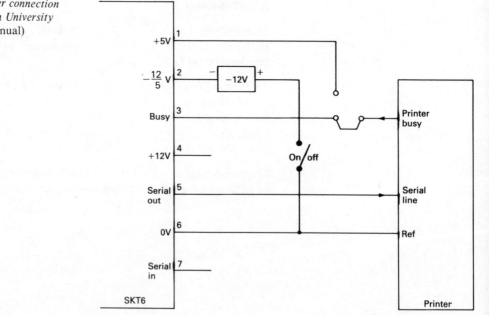

This interface is RS 232 (V24) compatible. This represents 0 and 1 by +12 V and –12 V levels. As these supplies are not available, they will have to be supplied externally. A PP3-type battery will be sufficient for the negative supply and most printers will operate satisfactorily if the +12 V is connected to +5 V.

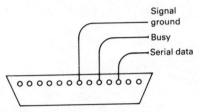

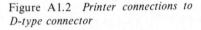

Signal ground
Busy
Serial data

Figure A1.2  *Printer connections to D-type connector*

If the BUSY signal is not available from your printer then this line should be tied to +5 V.

Figure A1.2 shows the connections to the standard 25-pin D-type connector. Check the connections to the 7-pin DIN plug as SK 6 does not conform in pin numbering with some DIN plugs. It is best to check on a pin-by-pin basis against Figure A1.3.

### Setting the baud rate

The serial data is software timed by the system. The constants for the baud rate are held at memory locations 2F41 and 2F42. The required values are given in Table A1.1.

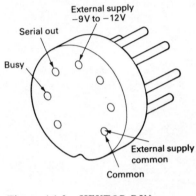

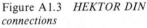

External supply
−9V to −12V
Serial out
Busy
External supply common
Common

Figure A1.3  *HEKTOR DIN connections*

**Table A1.1**  *Data for setting baud rate*

| Baud rate | 2F41 | 2F42 |
|-----------|------|------|
| 110 | 44 | 02 |
| 300 | D3 | 00 |
| 600 | 69 | 00 |
| 1200 | 33 | 00 |
| 2400 | 19 | 00 |
| 4800 | 0B | 00 |
| 9600 | 05 | 00 |

# Appendix 2  Data sheets

Material in Appendix 2 is reproduced by kind permission of Intel Corporation, Ferranti Electronics Limited and R. S. Components Limited.

# intel®

## 8251A
# PROGRAMMABLE COMMUNICATION INTERFACE

- **Synchronous and Asynchronous Operation**
  - **Synchronous:**
    5-8 Bit Characters
    Internal or External Character Synchronization
    Automatic Sync Insertion
  - **Asynchronous:**
    5-8 Bit Characters
    Clock Rate — 1, 16 or 64 Times Baud Rate
    Break Character Generation
    1, 1½, or 2 Stop Bits
    False Start Bit Detection
    Automatic Break Detect and Handling

- **Baud Rate — DC to 64k Baud**
- **Full Duplex, Double Buffered, Transmitter and Receiver**
- **Error Detection — Parity, Overrun, and Framing**
- **Fully Compatible with 8080/8085 CPU**
- **28-Pin DIP Package**
- **All Inputs and Outputs Are TTL Compatible**
- **Single 5 Volt Supply**
- **Single TTL Clock**

The 8251A is the enhanced version of the industry standard, Intel® 8251 Universal Synchronous/Asynchronous Receiver/Transmitter (USART), designed for data communications with Intel's new high performance family of microprocessors such as the 8085. The 8251A is used as a peripheral device and is programmed by the CPU to operate using virtually any serial data transmission technique presently in use (including IBM Bi-Sync). The USART accepts data characters from the CPU in parallel format and then converts them into a continuous serial data stream for transmission. Simultaneously, it can receive serial data streams and convert them into parallel data characters for the CPU. The USART will signal the CPU whenever it can accept a new character for transmission or whenever it has received a character for the CPU. The CPU can read the complete status of the USART at any time. These include data transmission errors and control signals such as SYNDET, TxEMPTY. The chip is constructed using N-channel silicon gate technology.

**PIN CONFIGURATION**          **BLOCK DIAGRAM**

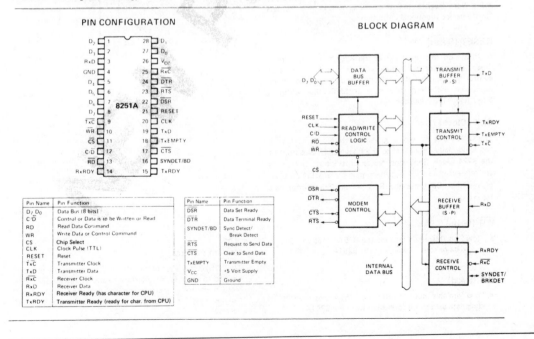

| Pin Name | Pin Function |
|----------|--------------|
| D₇–D₀ | Data Bus (8 bits) |
| C/D̄ | Control or Data is to be Written or Read |
| RD | Read Data Command |
| WR | Write Data or Control Command |
| CS | Chip Select |
| CLK | Clock Pulse (TTL) |
| RESET | Reset |
| TxC̄ | Transmitter Clock |
| TxD | Transmitter Data |
| RxC̄ | Receiver Clock |
| RxD | Receiver Data |
| RxRDY | Receiver Ready (has character for CPU) |
| TxRDY | Transmitter Ready (ready for char. from CPU) |

| Pin Name | Pin Function |
|----------|--------------|
| DSR | Data Set Ready |
| DTR | Data Terminal Ready |
| SYNDET/BD | Sync Detect/ Break Detect |
| RTS | Request to Send Data |
| CTS | Clear to Send Data |
| TxEMPTY | Transmitter Empty |
| Vcc | +5 Volt Supply |
| GND | Ground |

# 8251A

## 8251A BASIC FUNCTIONAL DESCRIPTION

### General

The 8251A is a Universal Synchronous/Asynchronous Receiver/Transmitter designed specifically for the 80/85 Microcomputer Systems. Like other I/O devices in a Microcomputer System, its functional configuration is programmed by the system's software for maximum flexibility. The 8251A can support virtually any serial data technique currently in use (including IBM "bi-sync").

In a communication environment an interface device must convert parallel format system data into serial format for transmission and convert incoming serial format data into parallel system data for reception. The interface device must also delete or insert bits or characters that are functionally unique to the communication technique. In essence, the interface should appear "transparent" to the CPU, a simple input or output of byte-oriented system data.

### Data Bus Buffer

This 3-state, bidirectional, 8-bit buffer is used to interface the 8251A to the system Data Bus. Data is transmitted or received by the buffer upon execution of INput or OUTput instructions of the CPU. Control words, Command words and Status information are also transferred through the Data Bus Buffer. The command status and data in, and data out are separate 8-bit registers to provide double buffering.

This functional block accepts inputs from the system Control bus and generates control signals for overall device operation. It contains the Control Word Register and Command Word Register that store the various control formats for the device functional definition.

### RESET (Reset)

A "high" on this input forces the 8251A into an "Idle" mode. The device will remain at "Idle" until a new set of control words is written into the 8251A to program its functional definition. Minimum RESET pulse width is 6 $t_{CY}$ (clock must be running).

### CLK (Clock)

The CLK input is used to generate internal device timing and is normally connected to the Phase 2 (TTL) output of the 8224 Clock Generator. No external inputs or outputs are referenced to CLK but the frequency of CLK must be greater than 30 times the Receiver or Transmitter data bit rates.

### $\overline{WR}$ (Write)

A "low" on this input informs the 8251A that the CPU is writing data or control words to the 8251A.

### $\overline{RD}$ (Read)

A "low" on this input informs the 8251A that the CPU is reading data or status information from the 8251A.

### C/$\overline{D}$ (Control/Data)

This input, in conjunction with the $\overline{WR}$ and $\overline{RD}$ inputs, informs the 8251A that the word on the Data Bus is either a data character, control word or status information.
1 = CONTROL/STATUS    0 = DATA

### $\overline{CS}$ (Chip Select)

A "low" on this input selects the 8251A. No reading or writing will occur unless the device is selected. When $\overline{CS}$ is high, the Data Bus in the float state and $\overline{RD}$ and $\overline{WR}$ will have no effect on the chip.

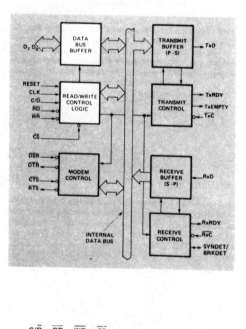

| C/$\overline{D}$ | $\overline{RD}$ | $\overline{WR}$ | $\overline{CS}$ | |
|---|---|---|---|---|
| 0 | 0 | 1 | 0 | 8251A DATA ⇒ DATA BUS |
| 0 | 1 | 0 | 0 | DATA BUS ⇒ 8251A DATA |
| 1 | 0 | 1 | 0 | STATUS ⇒ DATA BUS |
| 1 | 1 | 0 | 0 | DATA BUS ⇒ CONTROL |
| X | 1 | 1 | 0 | DATA BUS ⇒ 3-STATE |
| X | X | X | 1 | DATA BUS ⇒ 3-STATE |

### Modem Control

The 8251A has a set of control inputs and outputs that can be used to simplify the interface to almost any Modem. The Modem control signals are general purpose in nature and can be used for functions other than Modem control, if necessary.

## 8251A

### $\overline{DSR}$ (Data Set Ready)

The $\overline{DSR}$ input signal is a general purpose, 1-bit inverting input port. Its condition can be tested by the CPU using a Status Read operation. The $\overline{DSR}$ input is normally used to test Modem conditions such as Data Set Ready.

### $\overline{DTR}$ (Data Terminal Ready)

The $\overline{DTR}$ output signal is a general purpose, 1-bit inverting output port. It can be set "low" by programming the appropriate bit in the Command Instruction word. The $\overline{DTR}$ output signal is normally used for Modem control such as Data Terminal Ready or Rate Select.

### $\overline{RTS}$ (Request to Send)

The $\overline{RTS}$ output signal is a general purpose, 1-bit inverting output port. It can be set "low" by programming the appropriate bit in the Command Instruction word. The $\overline{RTS}$ output signal is normally used for Modem control such as Request to Send.

### $\overline{CTS}$ (Clear to Send)

A "low" on this input enables the 8251A to transmit serial data if the Tx Enable bit in the Command byte is set to a "one." If either a Tx Enable off or CTS off condition occurs while the Tx is in operation, the Tx will transmit all the data in the USART, written prior to Tx Disable command before shutting down.

### Transmitter Buffer

The Transmitter Buffer accepts parallel data from the Data Bus Buffer, converts it to a serial bit stream, inserts the appropriate characters or bits (based on the communication technique) and outputs a composite serial stream of data on the TxD output pin on the falling edge of $\overline{TxC}$. The transmitter will begin transmission upon being enabled if $\overline{CTS}$ = 0. The TxD line will be held in the marking state immediately upon a master Reset or when Tx Enable/$\overline{CTS}$ off or TxEMPTY.

### Transmitter Control

The transmitter Control manages all activities associated with the transmission of serial data. It accepts and issues signals both externally and internally to accomplish this function.

### TxRDY (Transmitter Ready)

This output signals the CPU that the transmitter is ready to accept a data character. The TxRDY output pin can be used as an interrupt to the system, since it is masked by Tx Disabled, or, for Polled operation, the CPU can check TxRDY using a Status Read operation. TxRDY is automatically reset by the leading edge of $\overline{WR}$ when a data character is loaded from the CPU.

Note that when using the Polled operation, the TxRDY status bit is *not* masked by Tx Enabled, but will only indicate the Empty/Full Status of the Tx Data Input Register.

### TxE (Transmitter Empty)

When the 8251A has no characters to transmit, the TxEMPTY output will go "high". It resets automatically upon receiving a character from the CPU. TxEMPTY can be used to indicate the end of a transmission mode, so that the CPU "knows" when to "turn the line around" in the half-duplexed operational mode. TxEMPTY is independent of the Tx Enable bit in the Command instruction.

In SYNChronous mode, a "high" on this output indicates that a character has not been loaded and the SYNC character or characters are about to be or are being transmitted automatically as "fillers". TxEMPTY does not go low when the SYNC characters are being shifted out.

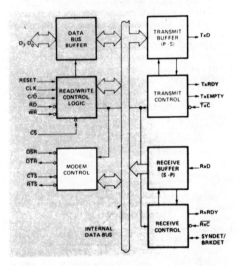

### $\overline{TxC}$ (Transmitter Clock)

The Transmitter Clock controls the rate at which the character is to be transmitted. In the Synchronous transmission mode, the Baud Rate (1x) is equal to the $\overline{TxC}$ frequency. In Asynchronous transmission mode the baud rate is a fraction of the actual $\overline{TxC}$ frequency. A portion of the mode instruction selects this factor; it can be 1, 1/16 or 1/64 the $\overline{TxC}$.

For Example:

    If Baud Rate equals 110 Baud,
    $\overline{TxC}$ equals 110 Hz (1x)
    $\overline{TxC}$ equals 1.76 kHz (16x)
    $\overline{TxC}$ equals 7.04 kHz (64x).

The falling edge of $\overline{TxC}$ shifts the serial data out of the 8251A.

## 8251A

### Receiver Buffer

The Receiver accepts serial data, converts this serial input to parallel format, checks for bits or characters that are unique to the communication technique and sends an "assembled" character to the CPU. Serial data is input to RxD pin, and is clocked in on the rising edge of $\overline{RxC}$.

### Receiver Control

This functional block manages all receiver-related activities which consist of the following features:

The RxD initialization circuit prevents the 8251A from mistaking an unused input line for an active low data line in the "break condition". Before starting to receive serial characters on the RxD line, a valid "1" must first be detected after a chip master Reset. Once this has been determined, a search for a valid low (Start bit) is enabled. This feature is only active in the asynchronous mode, and is only done once for each master Reset.

The False Start bit detection circuit prevents false starts due to a transient noise spike by first detecting the falling edge and then strobing the nominal center of the Start bit (RxD = low).

The Parity Toggle F/F and Parity Error F/F circuits are used for parity error detection and set the corresponding status bit.

The Framing Error Flag F/F is set if the Stop bit is absent at the end of the data byte (asynchronous mode), and also sets the corresponding status bit.

### RxRDY (Receiver Ready)

This output indicates that the 8251A contains a character that is ready to be input to the CPU. Rx RDY can be connected to the interrupt structure of the CPU or, for Polled operation, the CPU can check the condition of RxRDY using a Status Read operation.

Rx Enable off both masks and holds RxRDY in the Reset Condition. For Asynchronous mode, to set RxRDY, the Receiver must be Enabled to sense a Start Bit and a complete character must be assembled and transferred to the Data Output Register. For Synchronous mode, to set RxRDY, the Receiver must be enabled and a character must finish assembly and be transferred to the Data Output Register.

Failure to read the received character from the Rx Data Output Register prior to the assembly of the next Rx Data character will set overrun condition error and the previous character will be written over and lost. If the Rx Data is being read by the CPU when the internal transfer is occurring, overrun error will be set and the old character will be lost.

### $\overline{RxC}$ (Receiver Clock)

The Receiver Clock controls the rate at which the character is to be received. In Synchronous Mode, the Baud Rate (1x) is equal to the actual frequency of $\overline{RxC}$. In Asynchronous Mode, the Baud Rate is a fraction of the actual $\overline{RxC}$ frequency. A portion of the mode instruction selects this factor; 1, 1/16 or 1/64 the $\overline{RxC}$.

For Example:

Baud Rate equals 300 Baud, if
$\overline{RxC}$ equals 300 Hz (1x)
$\overline{RxC}$ equals 4800 Hz (16x)
$\overline{RxC}$ equals 19.2 kHz (64x).

Baud Rate equals 2400 Baud, if
$\overline{RxC}$ equals 2400 Hz (1x)
$\overline{RxC}$ equals 38.4 kHz (16x)
$\overline{RxC}$ equals 153.6 kHz (64x).

Data is sampled into the 8251A on the rising edge of $\overline{RxC}$.

**NOTE:** In most communications systems, the 8251A will be handling both the transmission and reception operations of a single link. Consequently, the Receive and Transmit Baud Rates will be the same. Both $\overline{TxC}$ and $\overline{RxC}$ will require identical frequencies for this operation and can be tied together and connected to a single frequency source ·(Baud Rate Generator) to simplify the interface.

### SYNDET (SYNC Detect)/BRKDET (Break Detect)

This pin is used in SYNChronous Mode for SYNDET and may be used as either input or output, programmable through the Control Word. It is reset to output mode low upon RESET. When used as an output (internal Sync mode), the SYNDET pin will go "high" to indicate that the 8251A has located the SYNC character in the Receive mode. If the 8251A is programmed to use double Sync characters (bisync), then SYNDET will go "high" in the middle of the last bit of the second Sync character. SYNDET is automatically reset upon a Status Read operation.

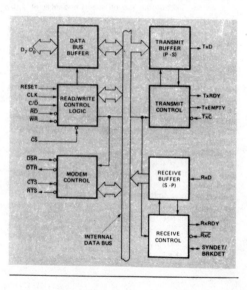

## 8251A

When used as an input (external SYNC detect mode), a positive going signal will cause the 8251A to start assembling data characters on the rising edge of the next $\overline{RxC}$. Once in SYNC, the "high" input signal can be removed. the period of $\overline{RxC}$. When External SYNC Detect is programmed, the Internal SYNC Detect is disabled.

### Break Detect (Async Mode Only)

This output will go high whenever an all zero word of the programmed length (including start bit, data bit, parity bit, and *one* stop bit) is received. Break Detect may also be read as a Status bit. It is reset only upon a master chip Reset or Rx Data returning to a "one" state.

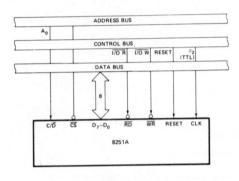

**8251A Interface to 8080 Standard System Bus**

## DETAILED OPERATION DESCRIPTION

### General

The complete functional definition of the 8251A is programmed by the system's software. A set of control words must be sent out by the CPU to initialize the 8251A to support the desired communications format. These control words will program the: BAUD RATE, CHARACTER LENGTH, NUMBER OF STOP BITS, SYNCHRONOUS or ASYNCHRONOUS OPERATION, EVEN/ODD/OFF PARITY, etc. In the Synchronous Mode, options are also provided to select either internal or external character synchronization.

Once programmed, the 8251A is ready to perform its communication functions. The TxRDY output is raised "high" to signal the CPU that the 8251A is ready to receive a data character from the CPU. This output (TxRDY) is reset automatically when the CPU writes a character into the 8251A. On the other hand, the 8251A receives serial data from the MODEM or I/O device. Upon receiving an entire character, the RxRDY output is raised "high" to signal the CPU that the 8251A has a complete character ready for the CPU to fetch. RxRDY is reset automatically upon the CPU data read operation.

The 8251A cannot begin transmission until the **Tx Enable** (Transmitter Enable) bit is set in the Command Instruction and it has received a Clear To Send ($\overline{CTS}$) input. The TxD output will be held in the marking state upon Reset.

### Programming the 8251A

Prior to starting data transmission or reception, the 8251A must be loaded with a set of control words generated by the CPU. These control signals define the complete functional definition of the 8251A and must immediately follow a Reset operation (internal or external).

The control words are split into two formats:

1. Mode Instruction
2. Command Instruction

### Mode Instruction

This format defines the general operational characteristics of the 8251A. It must follow a Reset operation (internal or external). Once the Mode Instruction has been written into the 8251A by the CPU, SYNC characters or Command Instructions may be inserted.

### Command Instruction

This format defines a status word that is used to control the actual operation of the 8251A.

Both the Mode and Command Instructions must conform to a specified sequence for proper device operation. The Mode Instruction must be inserted immediately following a Reset operation, prior to using the 8251A for data communication.

All control words written into the 8251A after the Mode Instruction will load the Command Instruction. Command Instructions can be written into the 8251A at any time in the data block during the operation of the 8251A. To return to the Mode Instruction format, the master Reset bit in the Command Instruction word can be set to initiate an internal Reset operation which automatically places the 8251A back into the Mode Instruction format. Command Instructions must follow the Mode Instructions or Sync characters.

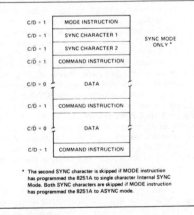

\* The second SYNC character is skipped if MODE instruction has programmed the 8251A to single character Internal SYNC Mode. Both SYNC characters are skipped if MODE instruction has programmed the 8251A to ASYNC mode.

**Typical Data Block**

## 8251A

### Mode Instruction Definition

The 8251A can be used for either Asynchronous or Synchronous data communication. To understand how the Mode Instruction defines the functional operation of the 8251A, the designer can best view the device as two separate components sharing the same package, one Asynchronous the other Synchronous. The format definition can be changed only after a master chip Reset. For explanation purposes the two formats will be isolated.

NOTE: When parity is enabled it is not considered as one of the data bits for the purpose of programming the word length. The actual parity bit received on the Rx Data line cannot be read on the Data Bus. In the case of a programmed character length of less than 8 bits, the least significant Data Bus bits will hold the data; unused bits are "don't care" when writing data to the 8251A, and will be "zeros" when reading the data from the 8251A.

### Asynchronous Mode (Transmission)

Whenever a data character is sent by the CPU the 8251A automatically adds a Start bit (low level) followed by the data bits (least significant bit first), and the programmed number of Stop bits to each character. Also, an even or odd Parity bit is inserted prior to the Stop bit(s), as defined by the Mode Instruction. The character is then transmitted as a serial data stream on the TxD output. The serial data is shifted out on the falling edge of $\overline{TxC}$ at a rate equal to 1, 1/16, or 1/64 that of the $\overline{TxC}$, as defined by the Mode Instruction. BREAK characters can be continuously sent to the TxD if commanded to do so.

When no data characters have been loaded into the 8251A the TxD output remains "high" (marking) unless a Break (continuously low) has been programmed.

### Asynchronous Mode (Receive)

The RxD line is normally high. A falling edge on this line triggers the beginning of a START bit. The validity of this START bit is checked by again strobing this bit at its nominal center (16X or 64X mode only). If a low is detected again, it is a valid START bit, and the bit counter will start counting. The bit counter thus locates the center of the data bits, the parity bit (if it exists) and the stop bits. If parity error occurs, the parity error flag is set. Data and parity bits are sampled on the RxD pin with the rising edge of $\overline{RxC}$. If a low level is detected as the STOP bit, the Framing Error flag will be set. The STOP bit signals the end of a character. Note that the *receiver* requires only *one* stop bit, regardless of the number of stop bits programmed. This character is then loaded into the parallel I/O buffer of the 8251A. The RxRDY pin is raised to signal the CPU that a character is ready to be fetched. If a previous character has not been fetched by the CPU, the present character replaces it in the I/O buffer, and the OVERRUN Error flag is raised (thus the previous character is lost). All of the error flags can be reset by an Error Reset Instruction. The occurrence of any of these errors will not affect the operation of the 8251A.

### Mode Instruction Format, Asynchronous Mode

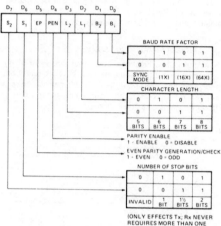

(ONLY EFFECTS Tx; Rx NEVER REQUIRES MORE THAN ONE STOP BIT)

### Asynchronous Mode

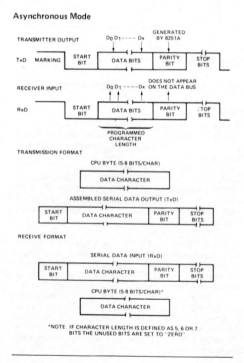

*NOTE: IF CHARACTER LENGTH IS DEFINED AS 5, 6 OR 7 BITS THE UNUSED BITS ARE SET TO "ZERO".

## 8251A

### Synchronous Mode (Transmission)

The TxD output is continuously high until the CPU sends its first character to the 8251A which usually is a SYNC character. When the $\overline{CTS}$ line goes low, the first character is serially transmitted out. All characters are shifted out on the falling edge of $\overline{TxC}$. Data is shifted out at the same rate as the $\overline{TxC}$.

Once transmission has started, the data stream at the TxD output must continue at the $\overline{TxC}$ rate. If the CPU does not provide the 8251A with a data character before the 8251A Transmitter Buffers become empty, the SYNC characters (or character if in single SYNC character mode) will be automatically inserted in the TxD data stream. In this case, the TxEMPTY pin is raised high to signal that the 8251A is empty and SYNC characters are being sent out. TxEMPTY does not go low when the SYNC is being shifted out (see figure below). The TxEMPTY pin is internally reset by a data character being written into the 8251A.

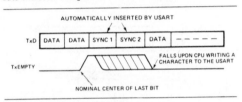

AUTOMATICALLY INSERTED BY USART

TxD | DATA | DATA | SYNC 1 | SYNC 2 | DATA | — — — —

TxEMPTY

FALLS UPON CPU WRITING A CHARACTER TO THE USART

NOMINAL CENTER OF LAST BIT

### Synchronous Mode (Receive)

In this mode, character synchronization can be internally or externally achieved. If the SYNC mode has been programmed, ENTER HUNT command should be included in the first command instruction word written. Data on the RxD pin is then sampled in on the rising edge of $\overline{RxC}$. The content of the Rx buffer is compared at every bit boundary with the first SYNC character until a match occurs. If the 8251A has been programmed for two SYNC characters, the subsequent received character is also compared; when both SYNC characters have been detected, the USART ends the HUNT mode and is in character synchronization. The SYNDET pin is then set high, and is reset automatically by a STATUS READ. If parity is programmed, SYNDET will not be set until the middle of the parity bit instead of the middle of the last data bit.

In the external SYNC mode, synchronization is achieved by applying a high level on the SYNDET pin, thus forcing the 8251A out of the HUNT mode. The high level can be removed after one $\overline{RxC}$ cycle. An ENTER HUNT command has no effect in the asynchronous mode of operation.

Parity error and overrun error are both checked in the same way as in the Asynchronous Rx mode. Parity is checked when not in Hunt, regardless of whether the Receiver is enabled or not.

The CPU can command the receiver to enter the HUNT mode if synchronization is lost. This will also set all the used character bits in the buffer to a "one", thus preventing a possible false SYNDET caused by data that happens to be in the Rx Buffer at ENTER HUNT time. Note that

the SYNDET F/F is reset at each Status Read, regardless of whether internal or external SYNC has been programmed. This does not cause the 8251A to return to the HUNT mode. When in SYNC mode, but not in HUNT, Sync Detection is still functional, but only occurs at the "known" word boundaries. Thus, if one Status Read indicates SYNDET and a second Status Read also indicates SYNDET, then the programmed SYNDET characters have been received since the previous Status Read. (If double character sync has been programmed, then both sync characters have been contiguously received to gate a SYNDET indication.) When external SYNDET mode is selected, internal Sync Detect is disabled, and the SYNDET F/F may be set at any bit boundary.

### Mode Instruction Format

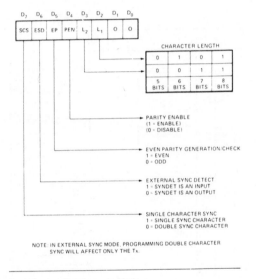

| $D_7$ | $D_6$ | $D_5$ | $D_4$ | $D_3$ | $D_2$ | $D_1$ | $D_0$ |
|---|---|---|---|---|---|---|---|
| SCS | ESD | EP | PEN | $L_2$ | $L_1$ | 0 | 0 |

CHARACTER LENGTH

| 0 | 1 | 0 | 1 |
|---|---|---|---|
| 0 | 0 | 1 | 1 |
| 5 BITS | 6 BITS | 7 BITS | 8 BITS |

PARITY ENABLE
(1 = ENABLE)
(0 = DISABLE)

EVEN PARITY GENERATION/CHECK
1 = EVEN
0 = ODD

EXTERNAL SYNC DETECT
1 = SYNDET IS AN INPUT
0 = SYNDET IS AN OUTPUT

SINGLE CHARACTER SYNC
1 = SINGLE SYNC CHARACTER
0 = DOUBLE SYNC CHARACTER

NOTE: IN EXTERNAL SYNC MODE, PROGRAMMING DOUBLE CHARACTER SYNC WILL AFFECT ONLY THE Tx.

### Data Format, Synchronous Mode

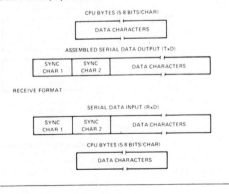

CPU BYTES (5-8 BITS/CHAR)

| DATA CHARACTERS |
|---|

ASSEMBLED SERIAL DATA OUTPUT (TxD)

| SYNC CHAR 1 | SYNC CHAR 2 | DATA CHARACTERS |
|---|---|---|

RECEIVE FORMAT

SERIAL DATA INPUT (RxD)

| SYNC CHAR 1 | SYNC CHAR 2 | DATA CHARACTERS |
|---|---|---|

CPU BYTES (5-8 BITS/CHAR)

| DATA CHARACTERS |
|---|

## 8251A

### COMMAND INSTRUCTION DEFINITION

Once the functional definition of the 8251A has been programmed by the Mode Instruction and the Sync Characters are loaded (if in Sync Mode) then the device is ready to be used for data communication. The Command Instruction controls the actual operation of the selected format. Functions such as: Enable Transmit/Receive, Error Reset and Modem Controls are provided by the Command Instruction.

Once the Mode Instruction has been written into the 8251A and Sync characters inserted, if necessary, then all further "control writes" (C/$\overline{D}$ = 1) will load a Command Instruction. A Reset Operation (internal or external) will return the 8251A to the Mode Instruction format.

### STATUS READ DEFINITION

In data communication systems it is often necessary to examine the "status" of the active device to ascertain if errors have occurred or other conditions that require the processor's attention. The 8251A has facilities that allow the programmer to "read" the status of the device at any time during the functional operation. (The status update is inhibited during status read).

A normal "read" command is issued by the CPU with C/$\overline{D}$ = 1 to accomplish this function.

Some of the bits in the Status Read Format have identical meanings to external output pins so that the 8251A can be used in a completely Polled environment or in an interrupt driven environment. TxRDY is an exception.

Note that status update can have a maximum delay of 28 clock periods from the actual event affecting the status.

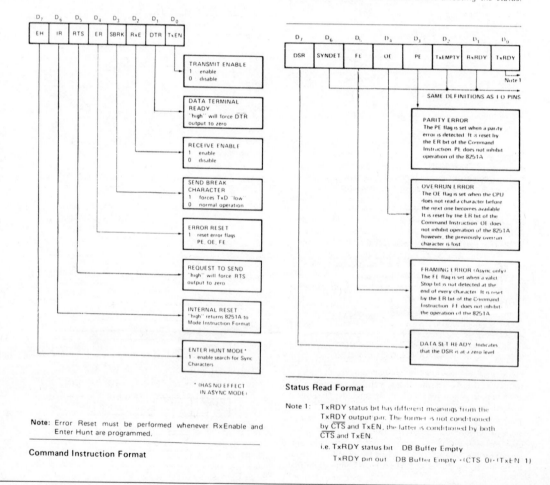

**Note:** Error Reset must be performed whenever RxEnable and Enter Hunt are programmed.

**Command Instruction Format**

**Status Read Format**

Note 1:   TxRDY status bit has different meanings from the TxRDY output pin. The former is not conditioned by $\overline{CTS}$ and TxEN, the latter is conditioned by both $\overline{CTS}$ and TxEN.

i.e. TxRDY status bit   DB Buffer Empty

TxRDY pin out   DB Buffer Empty ·($\overline{CTS}$ 0)·(TxEN 1)

# 8251A

## APPLICATIONS OF THE 8251A

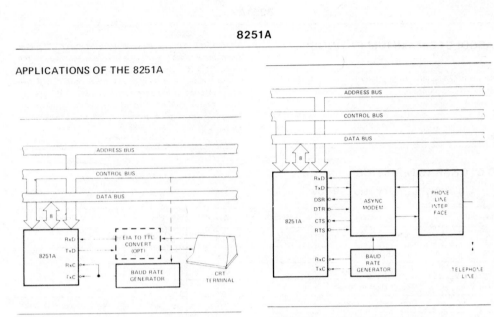

Asynchronous Serial Interface to CRT Terminal,
DC-9600 Baud

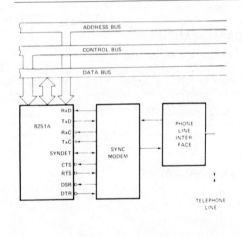

Asynchronous Interface to Telephone Lines

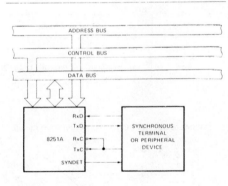

Synchronous Interface to Terminal or Peripheral Device

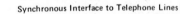

Synchronous Interface to Telephone Lines

## 8251A

### ABSOLUTE MAXIMUM RATINGS*

Ambient Temperature Under Bias. . . . . . . . 0°C to 70°C
Storage Temperature . . . . . . . . . . . . . . −65°C to +150°C
Voltage On Any Pin
    With Respect to Ground . . . . . . . . . . . −0.5V to +7V
Power Dissipation . . . . . . . . . . . . . . . . . . . . 1 Watt

*COMMENT: Stresses above those listed under "Absolute Maximum Ratings" may cause permanent damage to the device. This is a stress rating only and functional operation of the device at these or any other conditions above those indicated in the operational sections of this specification is not implied. Exposure to absolute maximum rating conditions for extended periods may affect device reliability.*

### D.C. CHARACTERISTICS

$T_A = 0°C$ to $70°C$; $V_{CC} = 5.0V \pm 5\%$; GND = 0V.

| Symbol | Parameter | Min. | Max. | Unit | Test Conditions |
|--------|-----------|------|------|------|-----------------|
| $V_{IL}$ | Input Low Voltage | −0.5 | 0.8 | V | |
| $V_{IH}$ | Input High Voltage | 2.0 | $V_{CC}$ | V | |
| $V_{OL}$ | Output Low Voltage | | 0.45 | V | $I_{OL} = 2.2$ mA |
| $V_{OH}$ | Output High Voltage | 2.4 | | V | $I_{OH} = -400\,\mu A$ |
| $I_{OFL}$ | Output Float Leakage | | ±10 | $\mu A$ | $V_{OUT} = V_{CC}$ TO 0.45V |
| $I_{IL}$ | Input Leakage | | ±10 | $\mu A$ | $V_{IN} = V_{CC}$ TO 0.45V |
| $I_{CC}$ | Power Supply Current | | 100 | mA | All Outputs = High |

### CAPACITANCE

$T_A = 25°C$; $V_{CC} = $ GND = 0V.

| Symbol | Parameter | Min. | Max. | Unit | Test Conditions |
|--------|-----------|------|------|------|-----------------|
| $C_{IN}$ | Input Capacitance | | 10 | pF | fc = 1MHz |
| $C_{I/O}$ | I/O Capacitance | | 20 | pF | Unmeasured pins returned to GND |

### TEST LOAD CIRCUIT:

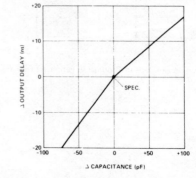

Figure 1.

TYPICAL Δ OUTPUT DELAY VS. Δ CAPACITANCE (dB)

## 8251A

### A.C. CHARACTERISTICS

$T_A = 0°C$ to $70°C$; $V_{CC} = 5.0V \pm 5\%$; GND = 0V

#### BUS PARAMETERS: (Note 1)

##### READ CYCLE

| SYMBOL | PARAMETER | MIN. | MAX. | UNIT | TEST CONDITIONS |
|---|---|---|---|---|---|
| $t_{AR}$ | Address Stable Before READ (CS, C/D) | 0 | | ns | Note 2 |
| $t_{RA}$ | Address Hold Time for READ (CS, C/D) | 0 | | ns | Note 2 |
| $t_{RR}$ | READ Pulse Width | 250 | | ns | |
| $t_{RD}$ | Data Delay from READ | | 200 | ns | 3, $C_L$ = 150 pF |
| $t_{DF}$ | READ to Data Floating | 10 | 100 | ns | |

##### WRITE CYCLE

| SYMBOL | PARAMETER | MIN. | MAX. | UNIT | TEST CONDITIONS |
|---|---|---|---|---|---|
| $t_{AW}$ | Address Stable Before WRITE | 0 | | ns | |
| $t_{WA}$ | Address Hold Time for WRITE | 0 | | ns | |
| $t_{WW}$ | WRITE Pulse Width | 250 | | ns | |
| $t_{DW}$ | Data Set Up Time for WRITE | 150 | | ns | |
| $t_{WD}$ | Data Hold Time for WRITE | 0 | | ns | |
| $t_{RV}$ | Recovery Time Between WRITES | 6 | | $t_{CY}$ | Note 4 |

NOTES: 1. AC timings measured $V_{OH}$ = 2.0, $V_{OL}$ = 0.8, and with load circuit of Figure 1.
2. Chip Select (CS) and Command/Data (C/D) are considered as Addresses.
3. Assumes that Address is valid before $\overline{RD}\downarrow$.
4. This recovery time is for Mode Initialization only. Write Data is allowed only when TxRDY = 1.
   Recovery Time between Writes for Asynchronous Mode is 8 $t_{CY}$ and for Synchronous Mode is 16 $t_{CY}$.

INPUT waveforms for AC tests:

## 8251A

OTHER TIMINGS:

| SYMBOL | PARAMETER | MIN. | MAX. | UNIT | TEST CONDITIONS |
|---|---|---|---|---|---|
| $t_{CY}$ | Clock Period | 320 | 1.35 | $\mu s$ | Notes 5, 6 |
| $t_\phi$ | Clock High Pulse Width | 120 | $t_{CY}-90$ | ns | |
| $t_{\bar\phi}$ | Clock Low Pulse Width | 90 | | ns | |
| $t_R$, $t_F$ | Clock Rise and Fall Time | 5 | 20 | ns | |
| $t_{DTx}$ | TxD Delay from Falling Edge of $\overline{TxC}$ | | 1 | $\mu s$ | |
| $t_{SRx}$ | Rx Data Set-Up Time to Sampling Pulse | 2 | | $\mu s$ | |
| $t_{HRx}$ | Rx Data Hold Time to Sampling Pulse | 2 | | $\mu s$ | |
| $f_{Tx}$ | Transmitter Input Clock Frequency | | | | |
| | 1x Baud Rate | DC | 64 | kHz | |
| | 16x Baud Rate | DC | 310 | kHz | |
| | 64x Baud Rate | DC | 615 | kHz | |
| $t_{TPW}$ | Transmitter Input Clock Pulse Width | | | | |
| | 1x Baud Rate | 12 | | $t_{CY}$ | |
| | 16x and 64x Baud Rate | 1 | | $t_{CY}$ | |
| $t_{TPD}$ | Transmitter Input Clock Pulse Delay | | | | |
| | 1x Baud Rate | 15 | | $t_{CY}$ | |
| | 16x and 64x Baud Rate | 3 | | $t_{CY}$ | |
| $f_{Rx}$ | Receiver Input Clock Frequency | | | | |
| | 1x Baud Rate | DC | 64 | kHz | |
| | 16x Baud Rate | DC | 310 | kHz | |
| | 64x Baud Rate | DC | 615 | kHz | |
| $t_{RPW}$ | Receiver Input Clock Pulse Width | | | | |
| | 1x Baud Rate | 12 | | $t_{CY}$ | |
| | 16x and 64x Baud Rate | 1 | | $t_{CY}$ | |
| $t_{RPD}$ | Receiver Input Clock Pulse Delay | | | | |
| | 1x Baud Rate | 15 | | $t_{CY}$ | |
| | 16x and 64x Baud Rate | 3 | | $t_{CY}$ | |
| $t_{TxRDY}$ | TxRDY Pin Delay from Center of last Bit | | 8 | $t_{CY}$ | Note 7 |
| $t_{TxRDY\ CLEAR}$ | TxRDY ↓ from Leading Edge of $\overline{WR}$ | | 150 | ns | Note 7 |
| $t_{RxRDY}$ | RxRDY Pin Delay from Center of last Bit | | 24 | $t_{CY}$ | Note 7 |
| $t_{RxRDY\ CLEAR}$ | RxRDY ↓ from Leading Edge of $\overline{RD}$ | | 150 | ns | Note 7 |
| $t_{IS}$ | Internal SYNDET Delay from Rising Edge of $\overline{RxC}$ | | 24 | $t_{CY}$ | Note 7 |
| $t_{ES}$ | External SYNDET Set-Up Time Before Falling Edge of $\overline{RxC}$ | | 16 | $t_{CY}$ | Note 7 |
| $t_{TxEMPTY}$ | TxEMPTY Delay from Center of Data Bit | | 20 | $t_{CY}$ | Note 7 |
| $t_{WC}$ | Control Delay from Rising Edge of WRITE (TxEn, $\overline{DTR}$, $\overline{RTS}$) | | 8 | $t_{CY}$ | Note 7 |
| $t_{CR}$ | Control to READ Set-Up Time ($\overline{DSR}$, $\overline{CTS}$) | | 20 | $t_{CY}$ | Note 7 |

5. The TxC and RxC frequencies have the following limitations with respect to CLK.
  For 1x Baud Rate , $f_{Tx}$ or $f_{Rx} \leq 1/(30\ t_{CY})$
  For 16x and 64x Baud Rate, $f_{Tx}$ or $f_{Rx} \leq 1/(4.5\ t_{CY})$

6. Reset Pulse Width = 6 $t_{CY}$ minimum; System Clock must be running during Reset.

7. Status update can have a maximum delay of 28 clock periods from the event affecting the status.

## 8251A

### SYSTEM CLOCK INPUT

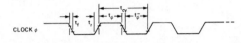

### TRANSMITTER CLOCK & DATA

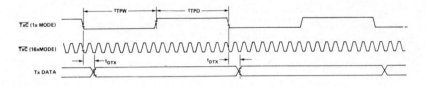

### RECEIVER CLOCK & DATA

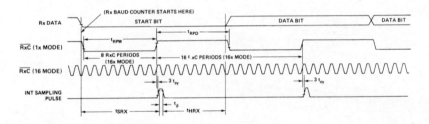

### WRITE DATA CYCLE (CPU → USART)

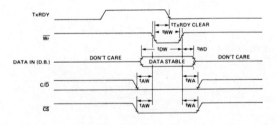

### READ DATA CYCLE (CPU ← USART)

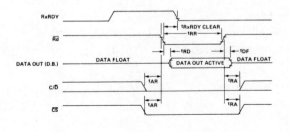

# 8251A

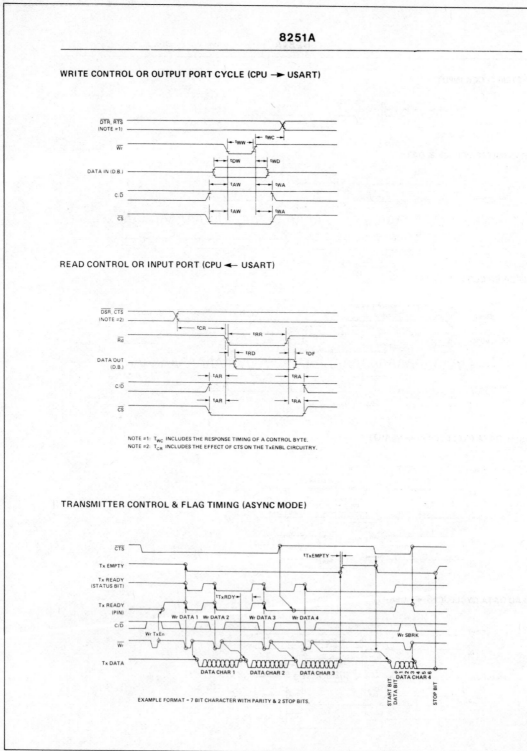

WRITE CONTROL OR OUTPUT PORT CYCLE (CPU → USART)

READ CONTROL OR INPUT PORT (CPU ← USART)

NOTE =1: $T_{WC}$ INCLUDES THE RESPONSE TIMING OF A CONTROL BYTE.
NOTE =2: $T_{CR}$ INCLUDES THE EFFECT OF CTS ON THE TxENBL CIRCUITRY.

TRANSMITTER CONTROL & FLAG TIMING (ASYNC MODE)

EXAMPLE FORMAT = 7 BIT CHARACTER WITH PARITY & 2 STOP BITS.

# 8251A

### RECEIVER CONTROL & FLAG TIMING (ASYNC MODE)

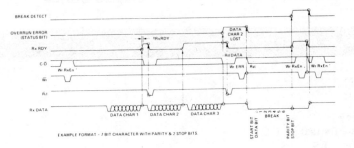

### TRANSMITTER CONTROL & FLAG TIMING (SYNC MODE)

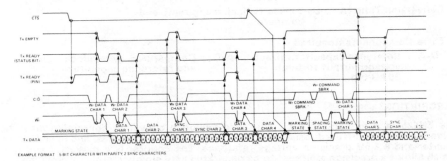

### RECEIVER CONTROL & FLAG TIMING (SYNC MODE)

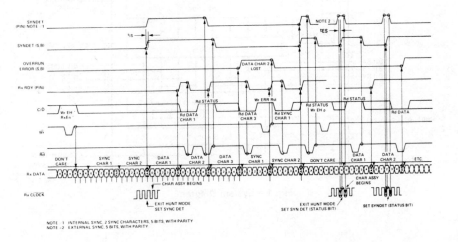

**D-A/A-D CONVERTER**

## ZN425 Series

### 8 Bit Monolithic D to A/A to D Converter

### FEATURES

- 8, 7 and 6 bit Accuracy
- 0°C to +70°C (ZN425E Series)
- −55°C to +125°C (ZN425J-8)
- TTL and 5V CMOS Compatible
- Single +5V Supply
- Settling Time (D to A) 1 μsec Typical
- Conversion Time (A to D) 1 msec typical, using ramp and
  compare.

- Extra Components Required

  D-A : Reference capacitor (direct      A-D : Comparator, gate, clock
  voltage output through                       and reference capacitor
  10 kΩ typ.)

### DESCRIPTION

The ZN425 is a monolithic 8-bit digital to analogue converter containing an R-2R ladder network of diffused resistors with precision bipolar switches, and in addition a counter and a 2.5V precision voltage reference. The counter is a powerful addition which allows a precision staircase to be generated very simply merely by clocking the counter.

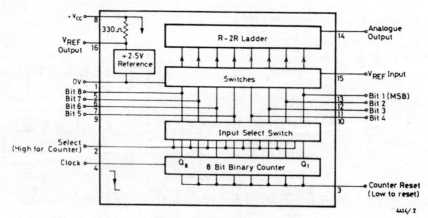

Fig. 1 – System Diagram

# ZN425 Series

## INTRODUCTION

The ZN425 is an 8-bit dual mode digital to analogue/analogue to digital converter. It contains an 8-bit D to A converter using an advanced design of R-2R ladder network and an array of precision bipolar switches plus an 8-bit binary counter and a 2.5 volt precision voltage reference all on a single monolithic chip.

The special design of ladder network results in full 8-bit accuracy using normal diffused resistors.

The use of the on-chip reference voltage is pin optional to retain flexibility. An external fixed or varying reference may therefore be substituted.

By including on the chip an 8-bit binary counter, analogue to digital conversion can be obtained simply by adding an external comparator (ZN424P) and clock inhibit gating (ZN7400E).

By simply clocking the counter the ZN425 can be used as a self-contained precision ramp generator.

A logic input select switch is incorporated which determines whether the precision switches accept the outputs from the binary counter or external digital inputs depending upon whether the control signal is respectively high or low.

The converter is of the voltage switching type and uses an R-2R resistor ladder network as shown in Fig. 2.

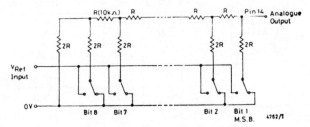

**Fig. 2 – The R-2R Ladder Network**

Each 2R element is connected either to 0V or $V_{REF}$ by transistor switches specially designed for low offset voltage (typically 1 millivolt).

Binary weighted voltages are produced at the output of the R-2R ladder, the value depending on the digital number applied to the bit inputs.

## ORDERING INFORMATION

| Operating Temperature | 8-bit Accuracy | 7-bit Accuracy | 6-bit Accuracy | Package |
|---|---|---|---|---|
| 0°C to +70°C | ZN425E-8 | ZN425E-7 | ZN425E-6 | Plastic |
| −55°C to +125°C | ZN425J-8 | — | — | Ceramic |

## ABSOLUTE MAXIMUM RATINGS

| | | |
|---|---|---|
| Supply voltage $V_{CC}$ .. .. .. | .. +7·0 volts | |
| Max. voltage, logic and $V_{REF}$ inputs .. | .. +5·5 volts *See note 3* | |
| Operating temperature range .. .. | .. 0°C to +70°C (ZN425E Series) | |
| | −55°C to +125°C (ZN425J-8) | |
| Storage temperature range .. .. | −55°C to +125°C | |

# ZN425 Series

CHARACTERISTICS (at $T_{amb}$ = 25°C and $V_{CC}$ = +5 volts unless otherwise specified).

*Internal voltage reference*

| Parameter | Symbol | Min. | Typ. | Max. | Units | Conditions |
|---|---|---|---|---|---|---|
| Output voltage | $V_{REF}$ | 2·4 | 2·55 | 2·7 | volts | I = 7·5 mA (internal) |
| Slope resistance | $R_s$ | — | 2 | 4 | ohms | I = 7·5 mA (internal) |
| $V_{REF}$ Temperature coefficient | — | — | 40 | — | ppm/°C | I = 7·5 mA (internal) |

*Note:* The internal reference requires a 0·22 µF stabilising capacitor between pins 1 and 16.

*8-Bit D to A Converter and Counter*

| Parameter | Symbol | Min. | Typ. | Max. | Units | Conditions |
|---|---|---|---|---|---|---|
| Resolution | | 8 | — | — | bits | |
| Accuracy (useful resolution)  ZN425J-8<br>ZN425E-8<br>ZN425E-7<br>ZN425E-6 | | 8<br>8<br>7<br>6 | —<br>—<br>—<br>— | —<br>—<br>—<br>— | bits<br>bits<br>bits<br>bits | $V_{REF}$ Input = 2 to 3V |
| Non-linearity | | — | — | ±0·5 | L.S.B. | *See Note 3* |
| Differential non-linearity | | — | ±0·5 | — | L.S.B. | *See Note 6* |
| Settling time | | — | 1·0 | — | µs | 1 L.S.B. step |
| Settling time to 0·5 L.S.B. | | — | 1·5 | 2·5 | µs | All bits ON to OFF or OFF to ON |
| Offset voltage  ZN425J-8<br>ZN425E-8⎫<br>ZN425E-6⎬<br>ZN425E-7⎭ | $V_{OS}$ | —<br><br>— | 8<br><br>3 | 12<br><br>8 | mV<br><br>mV | All bits OFF *See Note 3* |
| Full scale output | | 2·545 | 2·550 | 2·555 | volts | All bits ON Ext. $V_{REF}$=2·56V |
| Full scale temperature coeff. | | — | 3 | — | ppm/°C | Ext. $V_{REF}$=2·56V |
| Non-linearity error temp. coeff. | | — | 7·5 | — | ppm/°C | Relative to F.S.R. |
| Analogue output resistance | $R_o$ | — | 10 | — | kΩ | |
| External reference voltage | | 0 | — | 3·0 | volts | |
| Supply voltage | $V_{CC}$ | 4·5 | — | 5·5 | volts | *See Note 3* |
| Supply current | $I_s$ | — | 25 | 35 | mA | |
| High level input voltage | $V_{IH}$ | 2·0 | — | — | volts | *See Notes 1 and 2* |
| Low level input voltage | $V_{IL}$ | — | — | 0·7 | volts | |

# ZN425 Series

CHARACTERISTICS *(continued)*.

| Parameter | Symbol | Min. | Typ. | Max. | Units | Conditions |
|---|---|---|---|---|---|---|
| High level input current | $I_{IH}$ | — | — | 10 | μA | $V_{CC}$ = max.<br>$V_I$ = 2·4V |
| | | — | — | 100 | μA | $V_{CC}$ = max.<br>$V_I$ = 5·5V |
| Low level input current, bit inputs | $I_{IL}$ | — | — | −0·68 | mA | $V_{CC}$ = max.<br>$V_I$ = 0·3V |
| Low level input current, clock reset and input select | $I_L$ | — | — | −0·18 | mA | |
| High level output current | $I_{OH}$ | — | — | −40 | μA | |
| Low level output current | $I_{OL}$ | — | — | 1·6 | mA | |
| High level output voltage | $V_{OH}$ | 2·4 | — | — | volts | $V_{CC}$ = min.<br>Q = 1<br>$I_{load}$ = −40 μA |
| Low level output voltage | $V_{OL}$ | — | — | 0·4 | volts | $V_{CC}$ = min.<br>Q = 0<br>$I_{load}$ = 1·6 mA |
| Maximum counter clock frequency | $f_c$ | 3 | 5 | — | MHz | *See Note 5* |
| Reset pulse width | $t_R$ | 200 | — | — | ns | *See Note 4* |

*Notes:*

1. The Input Select pin (2) must be held low when the bit pins (5, 6, 7, 9, 10, 11, 12 and 13) are driven externally.

2. To obtain counter outputs on bit pins the Input Select pin (2) should be taken to $+V_{CC}$ via a 1 kΩ resistor.

3. The ZN425J differs from the ZN425E in the following respects:

    (a) For the ZN425J, the maximum linearity error may increase to ±1 LSB over the temperature ranges −55°C to 0°C and +70°C to +125°C.

    (b) Maximum operating voltage. Between 70°C and 125°C the maximum supply voltage is reduced to 5.0V.

    (c) Offset voltage. The difference is due to package lead resistance. This offset will normally be removed by the setting up procedure, and because the offset temperature coefficient is low, the specified accuracy will be maintained.

4. The device may be reset by gating from its own counter.

5. $F_{max}$ in A/D mode is 300 kHz, see page 1−18

6. Monotonic over full operating temperature range at resolution appropriate to accuracy.

# ZN425 Series

If Pin 2 is high then the output equals the Q output of the corresponding counter.

If Pin 2 is low then the output transistor, Tr1 is held off.

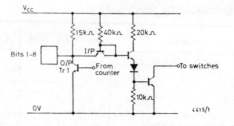

Fig. 3 – Bit Inputs/Outputs

## APPLICATIONS

### 1. 8-bit D to A Converter

The ZN425 gives an analogue voltage output directly from pin 14 therefore the usual current to voltage converting amplifier is not required. The output voltage drift, due to the temperature coefficient of the Analogue Output Resistance $R_o$, will be less than 0·004% per °C (or 1 L.S.B./100°C) if $R_L$ is chosen to be $\geqslant$ 650 kΩ.

In order to remove the offset voltage and to calibrate the converter a buffer amplifier is necessary. Fig. 4 shows a typical scheme using the internal reference voltage. To minimise temperature drift in this and similar applications the source resistance to the inverting input of the operational amplifier should be approximately 6 kΩ. The calibration procedure is as follows:

*i.* Set all bits to OFF (low) and adjust $R_2$ until $V_{out}$ = 0·000V.

*ii.* Set all bits to ON (high) and adjust $R_1$ until $V_{out}$ = Nominal full scale reading – 1 L.S.B.

*iii.* Repeat *i.* and *ii.*

$$\text{e.g. Set F.S.R. to } +3\cdot840 \text{ volts} - 1 \text{ L.S.B.}$$
$$= 3\cdot825 \text{ volts}$$
$$\left(1 \text{ L.S.B.} = \frac{3\cdot84}{256} = 15\cdot0 \text{ millivolts.}\right)$$

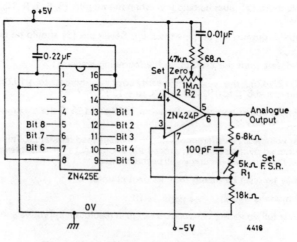

Fig. 4 – 8-bit Digital to Analogue Converter

# ZN425 Series

### 2. 8-bit Analogue to Digital Converter

A counter type ADC can be constructed by adding a voltage comparator and a latch as in Fig. 5. On the negative edge of the CONVERT COMMAND pulse (15 μs minimum) the counter is set to zero and the STATUS latch to logical 1. On the positive edge the gate is opened, enabling clock pulses to be fed to the counter input of the ZN425. The minimum negative clock pulse width to the ZN425 is 100 ns. The analogue output of the ZN425 ramps until it equals the voltage on the other input of the comparator. At this point the comparator output goes low and resets the STATUS latch to inhibit further clock pulses. The logical 0 from the status latch indicates that the 8 bit digital output is a valid representation of the analogue input voltage.

A small capacitor of 47 pF is added to the ZN425 output to stop any positive going glitches prematurely resetting the status latch. This capacitance is in parallel with the ZN425 output capacitance (20–30 pF) and they form a time constant with the ZN425 output resistance (10 kΩ). This time constant is the main limit to the maximum clock frequency. With a fast comparator the clock frequency can be up to 300 kHz. Using the ZN424P as a comparator the clock frequency should be restricted to 100 kHz. The conversion time varies with the input, being a maximum for full scale input.

$$\text{Maximum conversion time} = \frac{256}{\text{clock frequency in Hz}} \text{ seconds}$$

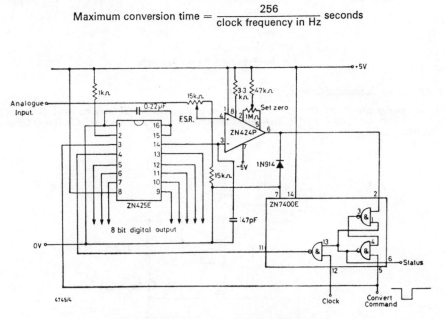

Fig. 5 – 8-bit Analogue to Digital Converter

### 3. Precision Ramp Generator

The inclusion of an 8-bit binary counter on the chip gives the ZN425 a useful ramp generator function. The circuit, Fig. 6 uses the same buffer stages as the D to A converter. The calibration procedure is also the same. Holding pin 2 low will set all bits to ON and if RESET is taken low with pin 2 high all the bits are turned OFF. If the end voltages of the ramp are not required to be set accurately then the buffer stage could be omitted and the voltage ramp will appear directly at pin 14.

# ZN425 Series

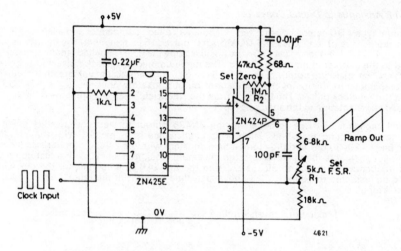

Fig. 6 – Precision Ramp Generator

### 4. Alternative Output Buffer using the ZLD741

The following circuit, employing the ZLD741 operational amplifier, may be used as the output buffer for both the 8-bit Digital to Analogue Converter (Fig. 4) and the Precision Ramp Generator (Fig. 6).

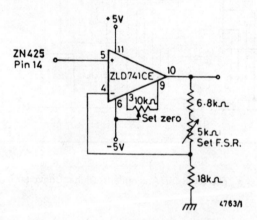

Fig. 7 – The ZLD741 as Output Buffer

### 5. Further Applications

Details of a wide range of additional applications, described in the Ferranti publication 'Application Report–ZN425 8-bit A-D/D-A Converter', are also available.

# ZN425 Series

## PIN CONNECTIONS

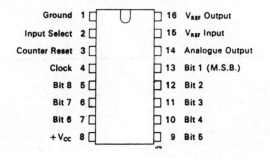

| | | | | |
|---|---|---|---|---|
| Ground | 1 | | 16 | $V_{REF}$ Output |
| Input Select | 2 | | 15 | $V_{REF}$ Input |
| Counter Reset | 3 | | 14 | Analogue Output |
| Clock | 4 | | 13 | Bit 1 (M.S.B.) |
| Bit 8 | 5 | | 12 | Bit 2 |
| Bit 7 | 6 | | 11 | Bit 3 |
| Bit 6 | 7 | | 10 | Bit 4 |
| +$V_{CC}$ | 8 | | 9 | Bit 5 |

## CHIP DIMENSIONS AND LAYOUT

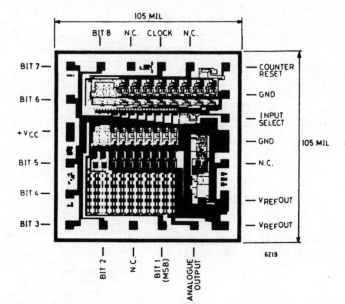

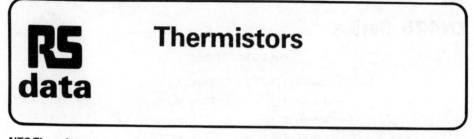

*Issued July 1983* **1867**

# Thermistors

## NTC Thermistors

The RS range of NTC thermistors includes standard tolerance negative temperature coefficient thermistors, plus a range of small close tolerance R/T curve matched thermistors.

## Standard Tolerance Thermistors

A range of 9 negative temperature coefficient bead thermistors constructed from a compound of nickel mangenite. Six types sealed in glass together with two stainless steel probe assemblies designed primarily for temperature measurement and control, flow measurement and liquid level detection. The remaining type, suspended in an evacuated glass envelope, is designed for applications in amplitude control, temperature compensation and time delay circuits.

| Characteristic Resistance | UNITS | Miniature beads | | | | Beads | | Probe ass. | | Evac type |
|---|---|---|---|---|---|---|---|---|---|---|
| | | 151-136 | 151-142 | 151-158 | 151-164 | 151-029 | 151-013 | 151-120 | 151-170 | 151-114 |
| $R_{BEAD}$  20°C | Ω | — | — | — | — | 2k | 1M | — | — | 5k |
| 25°C | Ω | 1k | 4.7k | 47k | 470k | — | — | 4.7k | 1.0M | — |
| $R_{MIN}$ (HOT) | Ω | 59 | 271 | 338 | 440 | 115 | 170 | 500 | 800 | 79 |
| $R_{BEAD}$ TOLERANCE | % | ±20 | ±20 | ±20 | ±20 | ±20 | ±20 | ±2 | ±2 | ±20 |
| $T_A$ max ambient temp range maximum dissipation | °C | −80 to +125 | −80 to +125 | −60 to +200 | −25 to +300 | −80 to +300 | −25 to +300 | −30 to +100 | −30 to +250 | 0 to +155 |
| Maximum dissipation | mW | 70 | 70 | 120 | 190 | 130 | 340 | 50 | 50 | 3·0 |
| Derate to zero at | °C | 125 | 125 | 200 | 300 | 125 | 300 | 100 | 250 | 225 |
| Dissipation constant | mW/°c | 0.7 | 0.7 | 0.7 | 0.7 | 1.2 | 1.2 | 5.0 | 5.0 | $12.5 \times 10^{-3}$ |
| Thermal time constant | s | 5 | 5 | 5 | 5 | 19 | 19 | 180 | 180 | 11 |
| B constant (25 to 100°C) | °K | 3000 | 3390 | 3980 | 4320 | 3200 | 4850 | 3275 | 5000 | 3250 |
| B tolerance | % | ±3 | ±3 | ±3 | ±3 | ±5 | ±5 | ±2 | ±2 | ±5 |
| Equivalent types | | GM102 VA3400 | GM472 VA3404 | GM473 VA3410 | GM474 | GL23 | GL16 | JA03 | JA09 | RA53 |

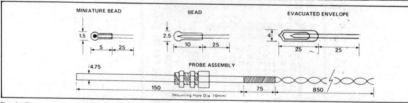

MINIATURE BEAD     BEAD     EVACUATED ENVELOPE

1.5     2.5     4

5   25     10   25     25   25

4.75     PROBE ASSEMBLY

150     (Mounting Hole Dia 10mm)     75     850

## Basic Formulae

The temperature coefficient $\alpha$ at any temperature within the operating range may be obtained from the formula:-

$$\alpha = -\frac{B}{T^2} \text{ (per °C)}$$

To determine the resistance at any temperature within the operating range may be obtained from the formula:-

$$R_2 = R_1 . e^{\left(\frac{B}{t_2} - \frac{B}{t_1}\right)}$$

where

B = characteristic temperature constant (°K)

T = bead temperature in (°K)

$R_1$ = resistance of thermistor at temperature $t_1(\Omega)$

$R_2$ = resistance of thermistor at temperature $t_2(\Omega)$

e = 2.7183

(Temperature in °K = temperature in °C + 273)

### Application Notes

Typical applications include temperature control of ovens, deep freezers, rooms and for process control, etc. Can also be used to drive high and low temperature alarms.

In the basic circuit below, calibration should be carried out by comparison with a known standard (e.g. a thermometer or thermocouple). In the case of 0°C a mixture of ice and water can be used and for 100°C use boiling water.

Note than non-linearity should be expected at extended temperatures.

1867

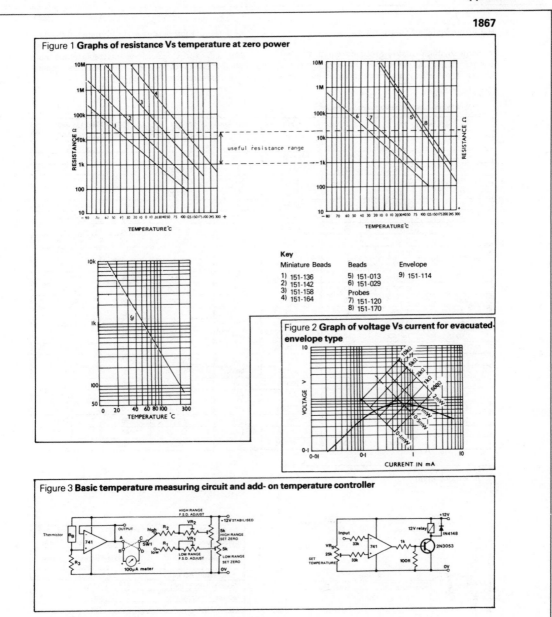

Figure 1 **Graphs of resistance Vs temperature at zero power**

useful resistance range

Key

Miniature Beads
1) 151-136
2) 151-142
3) 151-158
4) 151-164

Beads
5) 151-013
6) 151-029

Probes
7) 151-120
8) 151-170

Envelope
9) 151-114

Figure 2 **Graph of voltage Vs current for evacuated envelope type**

Figure 3 **Basic temperature measuring circuit and add- on temperature controller**

Table 1 Typical Resistor Values for Temperature Measuring Circuit (above)

| Thermistor | | Temperature in (°C) | | Resistor values (kΩ) | | | | |
|---|---|---|---|---|---|---|---|---|
| | | LOW | HIGH | R₁ | R₂ | R₃ | VR₁ | VR₂ |
| Miniature | 151-136 | 0 to −60∗ | 0 to 30 | 56 | 6.8 | 22 | 50 | 5 |
| | 151-142 | 0 to 30 | 0 to 100 | 18 | 33 | 22 | 10 | 20 |
| | 151-158 | 50 to 100 | 100 to 150 | 27 | 8.2 | 22 | 10 | 5 |
| | 151-164 | 150 to 200 | 200 to 250 | 12 | 3.9 | 22 | 5 | 2 |
| Beads | 151-029 | 0 to −30∗ | 0 to 30 | 27 | 10 | 22 | 20 | 5 |
| | 151-013 | 100 to 150 | 150 to 200 | 39 | 8.2 | 22 | 20 | 5 |
| Probe Assy. | 151-120 | 0 to −30∗ | 0 to 100 | 33 | 33 | 22 | 20 | 20 |

**NOTE**
∗ For negative ranges reverse meter by linking A to D and B to C

**1867**

## R-T Curve Matched Thermistors

A range of high quality precision curve matched thermistors, available in four characteristic resistances. They offer true interchangeability over a wide temperature range and eliminate the need for individual circuit adjustments or padding. These thermistors provide accurate and stable temperature sensing capability for applications such as temperature measurement and compensation.

**Table 2 Resistance/temperature characteristics**

| RS Stock No. **151-215** | | RS Stock No. **151-221** | | RS Stock No. **151-237** | | RS Stock No. **151-243** | |
|---|---|---|---|---|---|---|---|
| Temp °C | Res. Ω | Temp °C | Res. Ω | Temp °C | Res. Ω | Temp °C | Res. Ω |
| −80 | 2,210,400 | −80 | 3,684,000 | −80 | 7,368,000 | | |
| −70 | 935,250 | −70 | 1,558,800 | −70 | 3,117,500 | | |
| −60 | 421,470 | −60 | 702,450 | −60 | 1,404,900 | | |
| −50 | 201,030 | −50 | 335,050 | −50 | 670,100 | | |
| −40 | 100,950 | −40 | 168,250 | −40 | 336,500 | −40 | 4,015,500 |
| −30 | 53,100 | −30 | 88,500 | −30 | 177,000 | −30 | 2,064,000 |
| −20 | 29,121 | −20 | 48,535 | −20 | 97,070 | −20 | 1,103,400 |
| −10 | 16,599 | −10 | 27,665 | −10 | 55,330 | −10 | 611,870 |
| 0 | 9,795.0 | 0 | 16,325 | 0 | 32,650 | 0 | 351,020 |
| 10 | 5,970.0 | 10 | 9,950.0 | 10 | 19,900 | 10 | 207,850 |
| 20 | 3,747.0 | 20 | 6,245.0 | 20 | 12,490 | 20 | 126,740 |
| 25 | 3,000.0 | 25 | 5,000.0 | 25 | 10,000 | 25 | 100,000 |
| 30 | 2,417.1 | 30 | 4,028.5 | 30 | 8,057.0 | 30 | 79,422 |
| 40 | 1,598.1 | 40 | 2,663.3 | 40 | 5,327.0 | 40 | 51,048 |
| 50 | 1,080.9 | 50 | 1,801.5 | 50 | 3,603.0 | 50 | 33,591 |
| 60 | 746.40 | 60 | 1,244.0 | 60 | 2,488.0 | 60 | 22,590 |
| 70 | 525.60 | 70 | 876.00 | 70 | 1,752.0 | 70 | 15,502 |
| 80 | 376.50 | 80 | 627.50 | 80 | 1,255.0 | 80 | 10,837 |
| 90 | 274,59 | 90 | 457.65 | 90 | 915,30 | 90 | 7,707.7 |
| 100 | 203.49 | 100 | 339.15 | 100 | 678.30 | 100 | 5,569.3 |
| 110 | 153.09 | 110 | 255.15 | 110 | 510.30 | 110 | 4,082.9 |
| 120 | 116.79 | 120 | 194.65 | 120 | 389.30 | 120 | 3,033.3 |
| 130 | 90.279 | 130 | 150.47 | 130 | 300.93 | 130 | 2,281.0 |
| 140 | 70.581 | 140 | 117.64 | 140 | 235.27 | 140 | 1,734.3 |
| 150 | 55.791 | 150 | 92.985 | 150 | 185.97 | 150 | 1,331.9 |

### Dimensions (mm)

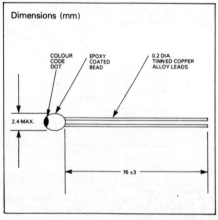

| RS Stock No | 151-215 | 151-221 | 151-237 | 151-243 |
|---|---|---|---|---|
| **Colour Code Dot** | Red | Orange | Yellow | Violet |
| Resistance at 20°C | 3kΩ | 5kΩ | 10kΩ | 100kΩ |
| Temperature Range | 0 to 70°C | | | |
| Tolerance (0 to 70°C) | ±0.2°C | | | |
| Dissipation Constant | 1mW | | | |
| Time Constant | 10s | | | |

### Definitions

*Dissipation constant.* Represents the amount of power required to raise the temperature of the thermistor. 1°C above its ambient temperature, expressed in milliwatts.

*Time constant.* The time required for the thermistor dissipating zero power is change 63% of the difference between its initial temperature value and that of a new impressed temperature environment.

## PTC Thermistors

The RS range of PTC thermistors includes three types for over-temperature protection and three types for over-current protection.

## Over-temperature Protection

A range of three positive temperature coefficient (PTC) thermistors, manufactured from a compound of barium lead and strontium titanates. The range consists of two disc types and one stud mounted version. These devices are primarily designed for temperature sensing and protection of semiconductor devices, transformers and motors etc. As can be seen from the resistance/temperature characteristic of fig. 6, the resistance of the PTC thermistor is low and relatively constant at low temperatures. As the ambient temperature increases, the resistance rises. The rate of increase becomes very rapid at the reference temperature (Tr) of the device. Tr is also known as the threshold, critical or switching temperature. Above Tr the characteristic becomes very steep and attains a high resistance value.

| Specification<br>RS Stock Nos. | | Stud<br>158-250 | Disc<br>158-266<br>158-272 |
|---|---|---|---|
| Maximum operating and storage temperature | | 155°C | Tr + 100°C |
| Minimum operating and storage temperature | | −20°C | −55°C |
| Typical thermal resistance (embedded) | (1) | — | 0.05°C/mW |
| Typical dissipation constant (embedded) | (1) | — | 20mW/°C |
| Maximum power dissipation at 25°C | (2) | — | 690mW |
| Maximum applied voltage at 25°C | (2) | — | 40V |
| Insulation between stud and lead | | 500V d.c. | — |
| Typical resistance at or below Tr − 20°C | | | 100Ω |
| Maximum resistance at or below Tr − 20°C | | | 250Ω |
| Maximum resistance at Tr − 5°C | (3) | | 550Ω |
| Typical resistance at Tr | (3) | | 1000Ω |
| Minimum resistance at Tr + 5°C | (3) | | 1330Ω |
| Minimum resistance at Tr + 15°C | (4) | | 4000Ω |

Notes
(1) Dependent on method of insulation and mounting
(2) Self heating in free air
(3) Measured at 2.5V d.c.
(4) Measured at 7.5V d.c.

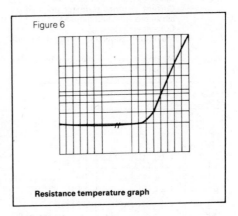

Figure 6

**Resistance temperature graph**

## Calibration

Calibration should be carried out by heating the thermistor to the appropriate reference temperature and adjust $R_2$ such that the appropriate L.E.D. lights.

## Series Connection

In temperature sensing circuits two or more devices may be connected in series. The sensing circuit will then indicate if any of the thermistors exceeds the reference temperature. An increase in the value of $R_1$ may be necessary to compensate for the additional volt drop across the thermistor.

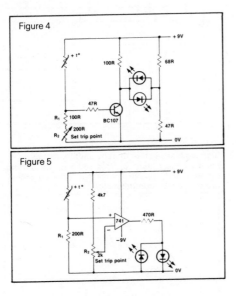

Figure 4

Figure 5

### Application Notes
Basic Temperature Sensing Circuits
Figure 4 shows a basic circuit which indicates when the temperature of the PTC thermistor is below Tr (i.e. safe operation) and will also indicate when Tr is exceeded. When both LEDs are off this indicates the Tr is being approached (approx. Tr −5°C).

Figure 5 shows a circuit which has a more defined 'trip point' than Figure 4 (set by $R_2$).

## Over-current Protection

Switching type Positive Temperature Coefficient (PTC) thermistors are prepared from compounds of barium, lead and strontium titanates to give a ceramic disc. Electrical contacts are made by the metallising of the disc faces using nickel, silver, etc; the completed disc is then provided with soldered lead wires.

## Definition of terms

$R_{min}$ —  Resistance of PTC at lowest point of R v T curve.

$R_{25}$ —  Resistance of PTC at 25°C.

$I_{max}$ —  Current value at turnover point of I/V curve at a specified temperature.

$I_{rest}$ —  Current value at $V_{max}$.

$I_{peak}$ —  Maximum allowable current through PTC.

$V_{max}$ —  Maximum voltage that may be applied to thermistor.

## Specification

Ratings

Resistance tolerance at 25°C ......................... ±25%

$V_{max}$ ............................................................. 265V rms

Ambient temperature range

Operating .................................................... 0-55°C

Storage ............................................. −40 to +155°C

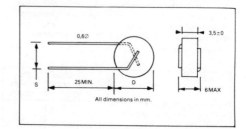

All dimensions in mm.

| RS stock No | Switch | $I_{max}$ (Typ.) | | $I_{REST}$ | $I_{PEAK}$ | Rs MIN | $(a < 1$ mA d.c. | | | Dimensions (mm) | |
| --- | --- | --- | --- | --- | --- | --- | --- | --- | --- | --- | --- |
| | | 25°C | 55°C | | | | $R_{25}$ | $R_{120}$ | $R_{155}$ | D | S |
| 151-287 | 125°C | 350 | 280 | <11,5 | 1,65A | 150Ω | 10 | < 30 | > 30k | 15 | 5 |
| 151-293 | | 200 | 155 | < 8,0 | 815mA | 300Ω | 25 | < 80 | > 80k | 10 | 5 |
| 151-300 | | 110 | 86 | < 7,0 | 250mA | 1kΩ | 70 | <218 | >220k | 6 | 5 |

Figure 7

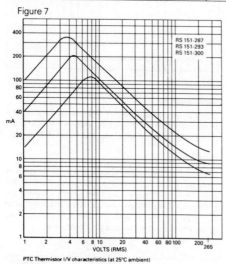

RS 151-287
RS 151-293
RS 151-300

PTC Thermistor I/V characteristics (at 25°C ambient)

Figure 8

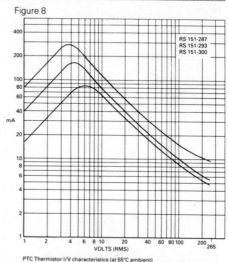

RS 151-287
RS 151-293
RS 151-300

PTC Thermistor I/V characteristics (at 55°C ambient)

## Theory of operation

The shape of the PTC thermistor resistance vs temperature characteristic, (Figure 9) can be considered in three distinct parts. The region from below 0°C to $R_{min}$ has a negative temperature coefficient of the order of 1%/°C; the region from $R_{min}$ to $R_{max}$ has a positive temperature coefficient in which values as high as 100%/°C can be realised. Beyond $R_{max}$ the TC is again negative. As a PTC Thermistor is sensitive to voltage variation, R v T curves are usually measured at a constant voltage. Figure 10 shows the characteristics of the load to be protected, together with the I/V of the thermistor on a linear scale. Region 'A' indicates the permissible load current range for normal operation. An increase in load current beyond the $I_{max}$ value will cause the thermistor to self-heat to a high resistance state thereby shifting its operating point to the region B. This reduces the current through and voltage across the load, effectively protecting the equipment etc.

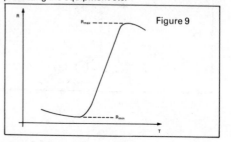

Figure 9

1867

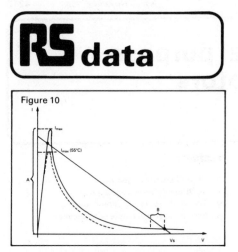

**RS data**

Figure 10

Similarly, if the ambient temperature surrounding the thermistor should, due to a fault condition, increase, the I/V curve will depress towards the dotted position. The load attempts to consume more than $I_{max}$ (55°C) and the thermistor will again self-heat and shift its operating point into the low current region.

## Selection

In order to ensure that the load is protected at the desired level and in the required reset mode, the following parameters must be taken into account:

1. Normal operating current range – Region 'A'.
2. 'Overload' current – $I_{max}$.
3. Operating temperature range – I/V curve shift.
4. Operating voltage range, (Vs).
5. Time response – position in Region 'A'.
6. Thermistor tolerances.
7. Permissible voltage drop across device.
8. Mounting arrangement.

The reset mode required, i.e. return to the 'A' region, is decided by the position of the load line in relation to the I/V curve. Figure 11 shows load line positions for the two modes, the auto-reset line intersects the I/V curve at only one point (F), thereby restricting stable operation to this point for normal load conditions. The manual or non-resetting line crosses the I/V curve at three

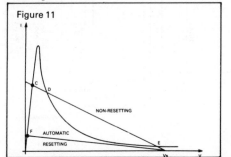

Figure 11

positions, giving the possibility of operation at either point. However, point D is in an unstable region so that in practice operation only occurs at points C or E.

If response time is a particularly important factor, the position of the operating point within region 'A', for a given device, (Figure 10) and the switch temperature of the PTC must be carefully considered. In circumstances where the circuit being protected is subject to short term overloads (which may be tolerated), the operating point should be the lower portion of region 'A'. Alternatively, where response time must be rapid, the operating point must be as close to the $I_{max}$ value as practicable, not forgetting the shift in characteristic with temperature.

Tolerances are usually quoted on the room temperature resistance (zero power), the higher values of $R_{25}$ giving the lower $I_{max}$.

As a thermistor is a resistive device there will inevitably be a voltage drop across it when in circuit. The maximum permissible voltage drop for the circuit concerned will dictate the room temperature or $R_{25}$ resistance value. It is usual to make the $R_{25}$ value in the order of 10% of the circuit resistance (or impedance).

The thermistor should be positioned in the equipment such that the surrounding air is reasonably still and unconfined. Moving air will effectively increase the $I_{max}$ value (at a given temperature) whilst confining the device will create a high ambient temperature, and therefore a lower $I_{max}$.

## Modification of I/V characteristics

In certain applications it is necessary to modify the I/V curve in order to produce the necessary characteristics. To obtain an auto-resetting device with a relatively high current rating, a resistor may be connected in parallel with the thermistor to 'lift' the characteristic to the dotted position, (Figure 12). This permits the load line to occupy a position in the upper 'A' region, but still crossing the combination curve at one stable point.

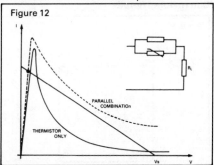

Figure 12

## Parallel operation

In instances where a sufficiently high $I_{max}$ value cannot be realised with existing devices, it is permissible to parallel connect two or more devices to achieve the required values; this may also be used to obtain lower $R_{25}$ resistances.

**R.S. Components Ltd.** 13-17 Epworth Street, London, EC2P 2HA  Telephone: 01-250 4000
For technical enquiries ring: 01-202 8252 or 0536-201234

An Electrocomponents Group Company  ©RS Components Ltd. 1982, 1983

Issued November 1980    **4210**

# RS data

# General purpose d.c. motors

A range of small d.c. motors with integral gearboxes for reliable yet economic usage in model systems, rotating displays, warning indicators, aerial drives and many other general purpose applications where rotational drive is required.

The motor consists of a permanent magnet stator with a three pole laminated iron rotor. A flat copper commutator has carbon brush contacts with a voltage dependent resistor, disc spark suppressor mounted between commutator and coils. This provides interference suppression and considerably increases brush life.

A steel ring completes the magnetic circuit and also provides a foundation for the plastic housing. Bearings are self-lubricating bronze and both reduction gears and housing are of tough polyacetal resin, which is highly resistant to chemicals and corrosion. Electrical connections are made via two solder tags on the motor body. Figure 1 gives motor dimensions.

## Features

- Combined motor-gearbox
- Integral interference suppressor
- Tough, corrosion-resistant housing
- Economic and versatile in application

## Quick reference table

| RS stock no. | 336-315 | 336-321 | 336-337 | 336-343 | symbol |
|---|---|---|---|---|---|
| nominal supply voltage | 12 | 12 | 6 | 6 | V d.c. |
| reduction ratio | 50:1 | 9:1 | 50:1 | 9:1 | |
| output speed | 60 | 330 | 60 | 330 | r.p.m. |
| output torque | 150 | 25 | 150 | 25 | mNm |

note: $1mNm = Nm \times 10^{-3} \simeq 10gcm$

## Dimensional details

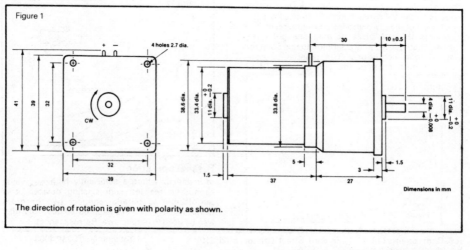

Figure 1

The direction of rotation is given with polarity as shown.

4210

 **General purpose d.c. motors**

## Typical motor characteristics  values apply to an ambient temperature of 22±5°C

| RS stock no. | 336-315 | 336-321 | 336-337 | 336-343 | symbol |
|---|---|---|---|---|---|
| nominal voltage | 12 | 12 | 6 | 6 | V d.c. |
| torque | 150 | 25 | 150 | 25 | mNm |
| speed at nominal load | 60 | 330 | 60 | 330 | r.p.m. |
| at no load | 78 | 415 | 78 | 415 | r.p.m. |
| current at nominal load | 185 | 185 | 360 | 360 | mA |
| at no load | 45 | 45 | 80 | 80 | mA |
| input power | 2.2 | 2.2 | 2.1 | 2.1 | W |
| Ambient temperature range | | −20 + 60 | | | °C |
| maximum axial play | | 0.5 | | | mm |

| Limiting conditions * | | | | | |
|---|---|---|---|---|---|
| max. input voltage | 18 | 18 | 9 | 9 | V d.c. |
| max. load | 150 | 37.5 | 150 | 37.5 | mNm |
| max. radial force on bearings | 6 | 2 | 6 | 2 | N |
| max. axial force | 6 | 2 | 6 | 2 | N |

## Typical curves

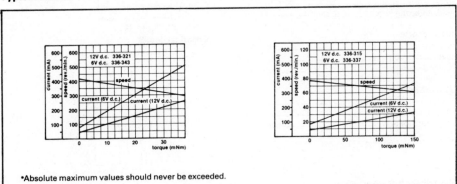

*Absolute maximum values should never be exceeded.

**R.S. Components Ltd.** 13-17 Epworth Street, London, EC2P 2HA  Telephone: 01-253 1222 or 01-250 4000
An Electrocomponents Group Company

*Issued July 1983* **4276**

# RS data

# Reflective and slotted opto switches

Gallium Arsenide infra-red emitting diodes and spectrally matched detectors housed in moulded packages mechanically designed to enable sensing in a variety of applications, i.e. limit switching, paper/tape sensing and optical encoding.

### Reflective opto switch
### Stock number 307-913
Comprises a Ga As infra-red emitting diode with a silicon phototransistor in a moulded rugged package. The sensor responds to the emitted radiation from the infra-red source only when a reflective object is within the field of view of the sensor. The device is ideal for such applications as end of tape detection, mark sensing, etc. An infra-red transmitting filter eliminates ambient illumination problems.

### Absolute maximum ratings at 25°C (unless stated)
Operating temp range —————— $-0°C$ to $70°C$
Storage temp range ——————— $-20°C$ to $80°C$
Lead soldering temperature (5 sec) ——— $260°C$

### Input diode
Forward d.c. current ————————— 40mA*
Reverse d.c. voltage ———————————— 2V
Power dissipation ————————————— 50mW**

### Output sensor
Collector – emitter voltage ——————— 15V
Emitter – collector voltage ——————— 5V
Power dissipation ————————————— 50mW**

  \* Derate linearly 0.73 mA/°C above 25°C
 \*\* Derate linearly 0.91 mW/°C above 25°C

## Applications
- Limit switch
- Paper sensor
- Counter
- Chopper
- Coin sensor
- Optical encoder
- Position sensor
- Level indicator

### Mechanical details

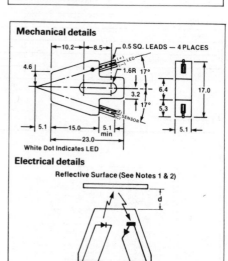

White Dot Indicates LED

### Electrical details

Reflective Surface (See Notes 1 & 2)

(+) (–)      (E) (C)
LED      Phototransistor Sensor

## Electrical characteristics
at 25°C (unless stated)

| Symbol | Parameter | min | typ | max | units | |
|---|---|---|---|---|---|---|
| **Input Diode** | | | | | | |
| $V_F$ | Forward Voltage | — | — | 1.8 | V | $I_F = 40mA$ |
| $I_R$ | Reverse Current | — | — | 100 | µA | $V_R = 2V$ |
| $P_O$ | Radiant Power | 0.5 | 1.5 | — | mW | $I_F = 20mA$ |
| **Output Sensor** | | | | | | |
| $BV_{CEO}$ | Collector-Emitter Breakdown Voltage | 15 | — | — | V | $I_{CE} = 100µA$ |
| $BV_{ECO}$ | Emitter-Collector Breakdown Voltage | 5 | — | — | V | $I_{EC} = 100µA$ |
| **Coupled** | | | | | | |
| $I_C$ | Photocurrent (see note 1) | 200 | — | — | µA | $I_F = 40mA$, $V_{CF} = 5V$ |
| $I_{CX}$ | Photocurrent (see note 2) | — | — | 20 | µA | $d = 5mm$ (see fig 2.) |

**4276**

## Typical characteristics

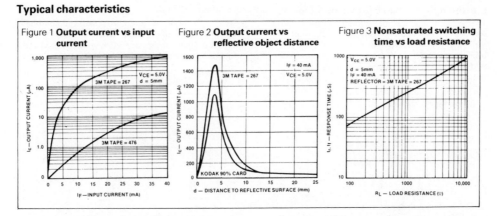

Figure 1 **Output current vs input current**

Figure 2 **Output current vs reflective object distance**

Figure 3 **Nonsaturated switching time vs load resistance**

Note 1: Photocurrent ($I_c$) is measured using 3M tape = 267 for a reflecting surface. The reflective qualities of 3M tape = 267 are very similar to an Eastman Kodak neutral white test card having 90% diffuse reflectance.

Note 2: Photocurrent ($I_{cx}$) is measured using 3M tape = 476 for a reflecting surface. 3M tape = 476 has a very black dull surface with optical reflectance qualities comparable to a surface coated with carbon black printers ink.

## Applications

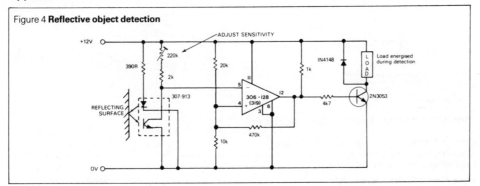

Figure 4 **Reflective object detection**

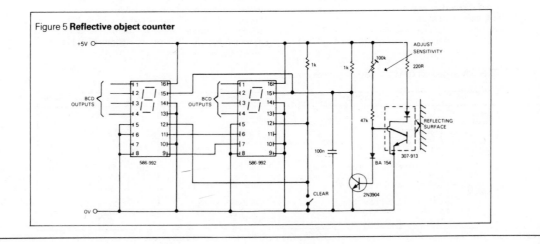

Figure 5 **Reflective object counter**

4276

### Slotted opto switches (Stock numbers 306-061, 304-560)

Two versions are available. 306-061 comprises a Ga As infra-red LED coupled with an npn silicon photo-transistor housed in a plastic package with infra-red transmitting filter for high ambient light application and dust protection. 304-560 is a similar device but the detector is an integrated circuit consisting of a Schmitt trigger, voltage regulator, differential amplifier and photodiode. The on-chip voltage regulator gives a wide operating voltage range and ensures output compatibility with TTL/LSTTL/CMOS logic.

Figure 6 **Mechanical details**

306-061                 304-560

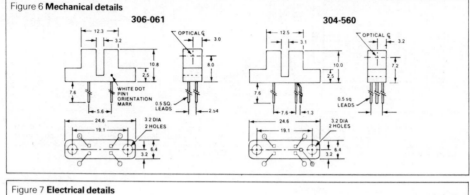

Figure 7 **Electrical details**

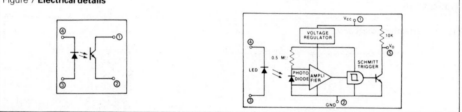

### Absolute maximum ratings at 25°C (unless stated)

|  | 306-061 | 304-560 |
|---|---|---|
| Operating temperature range | −55°C to 100°C | −40°C to 100°C |
| Storage temperature range | −55°C to 125°C | −55°C to 115°C |
| Lead soldering temperature (10s) | 260°C | 260°C |

**Input diode (306-061 and 304-560)**

| | |
|---|---|
| Forward d.c. current | 50mA |
| Peak forward current | 3A |
| (1μs p.w. 300pps) | |
| Reverse d.c. voltage | 3V |
| Power dissipation | 100mW |

* Derate linearly 1.33mW/°C above 25°C

**Output sensors**

|  | 306-061 | 304-560 |
|---|---|---|
| Collector – emitter voltage | 30V | — |
| Emitter – collector voltage | 5V | — |
| Max allowable $V_{CC}$ | — | 20V |
| Collector d.c. current | 30mA | 50mA |
| Power dissipation | 150mW** | 250mW |

Derate linearly 3.3mW/°C above 25°C

### Electrical characteristics at 25°C (unless stated)

| Symbol | Parameter | Conditions | min | typ | max | units |
|---|---|---|---|---|---|---|
| **Input Diode** | | | | | | |
| $V_F$ | Forward Voltage | $I_F = 20mA$ | | 1.2 | 1.7 | V |
| $I_R$ | Reverse Current | $V_R = 3V$ | | | 100 | μA |
| **Output Sensor** | | | | | | |
| $BV_{CEO}$ | Collector-Emitter Breakdown Voltage | $I_C = 1.0mA$ | 30 | 60 | | V |
| $BV_{ECO}$ | Emitter-Collector Breakdown Voltage | $I_E = 100μA$ | 5 | 8 | | V |
| $I_D$ | Collector Dark Current | $V_{CE} = 10V, I_F = O, H = O$ | | 10 | 100 | nA |
| **Coupled** | | | | | | |
| $V_{CE(SAT)}$ | Collector-Emitter Sat. Voltage | $I_F = 10mA, I_C = 250μA$ | | 0.2 | 0.4 | V |
| $I_{C(ON)}$ | On-state Collector Current | $I_F = 10mA, V_{CE} = 5V$ | 1000 | 3000 | | μA |
| $t_R$ | Response Time | | | 5 | | μS |

4276

# RS data

## Figure 8 On-state collector current vs input diode forward current

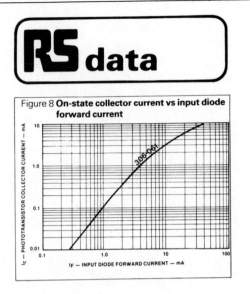

## Figure 9 Application: Event counting or limit switching

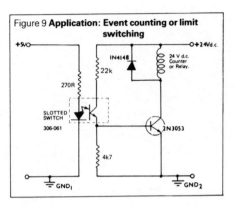

## 304-560 Opto Schmitt switch

| Symbol | Parameter | Conditions | min | typ | max | units |
|---|---|---|---|---|---|---|
| **Input Diode** $V_F$ | Forward Voltage | $I_F = 20$mA | | | 1.5 | V |
| $I_R$ | Reverse Current | $V_R = 3.0$V | | | 10 | $\mu$A |
| **Output Sensor** $V_{CC}$ | Operating supply voltage range | | 4.5 | | 16 | V |
| | Output Voltage (Low) | $-40°C < T_A < 100°C$. $I_O = 16$mA | | | 0.4 | V |
| | Output Voltage (High) | NB. Output tied to $V_{CC}$ through 10K resistor | | $V_{CC}$ | | |
| $I_{CC}$ | Operating Current | $V_{CC} = 16$V | | | 15 | mA |
| $t_p$ | Propagation delay time | $I_F = 10$mA | 1 | | 5 | $\mu$s |
| $t_r$ | Output rise time | $C_L = 50$pF $R_L = 390$R $V_{CC} = 5$V | | 150 | 180 | nS |
| $t_f$ | Output fall time | $C_L = 50$pF, $R_L = 390$R- $V_{CC} = 5$V | | 23 | 50 | nS |
| — | Hysteresis | Note 2 | 10 | | 30 | % |
| $I_{FT}$ | Required LED Current | Note 1. $-40°C < T_A < 75°C$ | | | 10 | mA |
| $f_{max}$ | Maximum operating frequency | $C_L = 50$pF, $R_L = 390$R $V_{CC} = 5$V | | | 100 | kHz |

Note 1: Required LED current is the minimum forward LED current required to trigger the detector output from LOW to HIGH. Higher LED current may be required for application where optical transmission is reduced.

Note 2: Hysteresis is defined in terms of irradiance (mW/cm$^2$) transmitted to the detector and is equal to the difference in the threshold point (min. irradiance to switch the output high) to the release point (reduced amount of irradiance to switch the output back low) divided by the threshold point.

**R.S. Components Ltd.** 13-17 Epworth Street, London, EC2P 2HA
For technical enquiries ring: 01-202 8252 or 0536-201234

Telephone: 01-250 4000

An Electrocomponents Group Company

©RS Components Ltd. 1982, 1983

# Index